Praise for *Stealing th...*

'This tragic tale of two daredevil surfers w...
...n't big enough for both of them is...

...prose perfectly befits his subject ... Martin's narrative, part factual, part fanciful, sweeps the reader along like one of the awesome incoming walls of water at Waimea Bay. A fascinating glimpse into obsession' Mike Rowbottom, *Independent*

'Something as compelling and carefully researched as *Stealing the Wave* would stand out in any field ... it grabs the attention like a ten-foot barrel on a two-foot day' Roger Cox, *Scotsman*

'Gripping ... Surfing emerges as a dangerous, solitary and potentially fatal obsession' *Daily Telegraph*

'An entertaining glimpse into both the seamy and sublime sides of surfing' *Sunday Telegraph*

'Captivating' *Esquire*

'A gripping insight into the mythic tradition and changing commercial face of a world where, as Foo would find out to his cost, life's a beach and then you die' Alan Chadwick, *Metro*

'The war between Ken Bradshaw and Mark Foo, as told in *Stealing the Wave* makes charging the mound seem tame by comparison. And as Martin pushes toward the story's tragic finale he manages to make the two men seem like they're battling not only for the hearts and minds of Brazilian groupies but for the fate of big-wave surfing itself' *New York Times*

'Martin brings both energy and skill to the evocation of the battle ... A dramatic story and one that Martin captures with understated grace' *Sydney Morning Herald*

BEWARE INVISIBLE COWS

MY SEARCH FOR THE SOUL OF THE UNIVERSE

ANDY MARTIN

SIMON &
SCHUSTER

London · New York · Sydney · Toronto

A CBS COMPANY

First published in Great Britain in 2009
by Simon & Schuster UK Ltd
A CBS COMPANY

The right of Harry Martin to be identified as the author of this work
has been asserted by him in accordance with sections 77
and 78 of the Copyright, Designs and Patents Act, 1988.

1 3 5 7 9 10 8 6 4 2

Simon & Schuster UK Ltd
1st Floor
222 Gray's Inn Road
London
WC1X 8HB

www.simonandschuster.co.uk

Simon & Schuster Australia
Sydney

PICTURE CREDITS
p.8: Bishop Museum/Keck Observatory
p.14: Peter Tuthill/Palomar/Keck Observatory
p.30: ESA, NASA, J. P. Kneib (Caltech/Observatoire Midi-Pyrénées)
and R. Ellis (Caltech)
p.44: M. Reck and P. G. Kwiat
p.98: NASA
p.116, 128: Weber, *Physical Review Letters*
p.190: Peter Saville
p.242: Emilio Segrè Visual Archives

A CIP catalogue for this book
is available from the British Library.

ISBN: 978-1-84737-416-5

Typeset by M Rules
Printed in the UK by CPI Mackays, Chatham ME5 8TD

For Heather, Spencer and Jack,
even if you didn't really believe in the UFO

I saw the inconceivable universe
– Jorge Luis Borges

ACKNOWLEDGEMENTS

If you want to see further, it definitely helps to stand on the shoulders of giants, but it is even more effective when they are armed with a decent telescope or an interferometer. I wouldn't have seen the Red Dot or the Red Square without the whole team at the Keck Observatory in Hawaii, up on Mauna Kea and down in Waimea. In this book I have made particular use of the thoughts and remarks of Bill Healey, Craig, Neil, Mike the IceMan, Tomas Krasczinski, and Chris the welder. I am grateful to Doug at the Visitor Information Station for introducing me to the wekiu bug. And I have fond if slightly blurry memories of a brief encounter one night with Ms A. Martin at the Waipio gas station.

Everyone at LIGO in Hanford was extremely tolerant of my curiosity and endless questions. I learned a lot from Fred Raab, Evan the grad student, Gregory Mandell, and Mike Landry, who kindly allowed me to make much more extraneous noise than he would normally put up with. Thanks to Jill Perry and Dale Ingram for getting me through the door. John on the exercise bike at the Shilo Inn in Richland told me about the high density of PhDs in the area. I am indebted to the kind people at the Center for Astronomical Research of Northwestern University, especially Marina, for setting me on the right track when I was lost and on the verge of a nervous breakdown.

At Caltech, in a slightly miraculous echo of the Creation, they made time and space for me. I had enlightening conversations with Kip Thorne, Alan Weinstein, Sterl Phinney, Albert Lazzarini, and Stuart Anderson. In the same neighbourhood I picked up useful clues from the Huntington Museum and the Mount Wilson Observatory.

In Portland, Oregon, I benefited from the hospitality of Jenny, Andrew, Daniel, and Nick.

In New York, the mysterious Mr M. provided one line of enquiry, but I am especially grateful to Avis Lang of the Hayden Planetarium for blowing most of my ideas out of the water.

In Cambridge I was inspired by the Astronomical Observatory. Sir Martin Rees, Astronomer Royal, provided some valuable hints and pointers on waves and Sandu Popescu helpfully explained how to transmit the soul of a particle. Sidney and Marion Abrahams were always encouraging. And Juliet Burridge looked through some telescopes with me.

David Marquis sent me two significant words from Canada.

Cambridge University let me go off at a tangent and, at the Center for Historical Analysis at Rutgers University all the participants in the 'Question of the West' seminar, and especially Jackson Lears, kept me going. David Godwin picked up the alien signals and Mike Jones turned them into a terrestrial book.

My big brother Charles was a great support throughout.

Heather, Jack and Spencer were my indispensable collaborators, as always.

Finally, I couldn't have made head or tail of half of this without all the input, over many years, from Unc, my rocket-scientist twin brother, who invented the 'farcom' and, even if he didn't invent them, made very good use of a couple of tin cans and a piece of string.

0

The thing I really wanted to know was: what happened to the *real* Andy Martin? It was a simple question of genealogy, but it seemed as if the branches of my family tree were outstandingly tangled, crooked, and gnarly.

According to mysterious phone calls from New York, I was not, in fact, Andy Martin, but rather the product of an ill-starred union between the scion of a wealthy American philanthropic family and the wife of the then Shah of Iran. For complicated and melodramatic reasons, apparently, we had been switched at birth, and I had been brought up as the child of Londoners in Essex, England, while the real Andy Martin was, presumably, swanning around somewhere in the United States or possibly the Middle East. Or could he have been the victim of an assassination attempt? That, I imagined, was the crux of why we had been swapped: a cunning security ruse. The common-or-garden Andy Martin (a mere cockney ragamuffin, after all) could be sacrificed, while I, a sleeper, the true – but yet to be unveiled and duly revered – prince and inheritor of all the riches of the Orient,

1

infinitely more exotic and important, would survive unscathed to emerge as the leader of – well, what precisely: the Iranian counter-revolution? The trouble with this Hollywood scenario of my origin was that I was as likely to be eliminated now as before, either as a subversive threat to the new regime or as an embarrassment to the old one.

The other problem, when I stopped to think about it clearly, was that the whole fabulous tale was exceedingly unlikely. The source of the phone calls was not some unimpeachable objective source but, in truth, the scion of that wealthy American philanthropic family himself. He revealed, as the calls continued, that he was currently undergoing treatment in a 'clinic' in New York, almost certainly of the psychiatric kind, possibly – I guessed – involving some kind of Walter Mitty syndrome. He was persistent. He kept on phoning – over a period of several months – and coming up with almost convincing evidence. He knew more than was reasonable to know about my 'adoptive' family and me. There was supposed to be a birth certificate somewhere, which he promised to FedEx over (I never saw it). There was some question of a cheque too (also elusive). But he was most likely deluded in his beliefs. Or this was a massive hoax of some sort. Pure soap opera. I was not taken in for a moment.

All the same, I couldn't help appealing to the mirror on the wall, from time to time, for faint resemblances to the famously beautiful widow of the deposed Shah and Ava Gardner lookalike, the Empress Soraya. Wasn't there, I started to tell myself, a definite Middle Eastern curve to those Anglo-Saxon features? I wrapped a towel around my head: wouldn't I look rather fetching in robes? I could be another Lawrence of Arabia, a modern-day Peter O'Toole, but with real blood ties to the East. Even now, when 'my' wealthy family donates another billion or so to Cambridge (or wherever), or I am passing through an art gallery bearing our name, I can't help but think: what about *me*? Isn't some part of that large sum properly *mine*? Wasn't there, after all, that tenuous and yet compelling piece of evidence that I had in fact adopted the family's name as an alias and nom de plume while a student, long before I even 'knew' of their existence? I could have chosen any name under the sun, but, no, I adopted my own secret name. Wasn't that a clincher?

It was a beautiful dream. I imagine that everyone, at one time or another, becomes convinced that they are, in fact, a long-lost prince or princess, somehow disguised – temporarily – as an ordinary citizen and ne'er-do-well. And that the moment will soon come when the truth will be revealed and we will all throw off our humble rags and stand forth in a beam of radiant light as the champion and defender of the kingdom. It is the stuff of myth and legend. In my own case, however, the chances of this hypothesis being right are less than – I would estimate – one-tenth of one per cent (and that is an optimistic exaggeration). To begin with, I look too much like my mother (not the wife of the Shah of Iran). On the other hand, it is true, I bear little resemblance to my twin brother.

This is the point that my mysterious caller kept zeroing in on, as if it proved his point. 'Why do you think you look absolutely nothing like your twin brother? It's obvious, isn't it: because you are *not* his brother at all!' QED. It was a tempting thought. My girlfriend tended to assent: 'I've often wondered how you two came to be brothers. You've got to admit – this would explain a lot.' Unc (as I called him for obscure and ancient reasons) was a scientist in Oxford, a specialist in X-ray astronomy, and had worked at the European Space Agency, whereas I was a research fellow in French at Cambridge. He had dark hair and brown eyes; I had fair hair and blue eyes. We were – so far as I knew, according to received wisdom – dizygotic (or 'fraternal'), emerging from completely separate ova, but I preferred to think of us as 'antithetical' rather than merely non-identical twins: radically opposed on nearly everything, totally out of sync, two distinct beings who happened to share the same birthday.

He was born before me. Either astrology was radically misleading or a lot of planets would have had to shuffle around in the ten minutes dividing us. By the time I arrived he had already established himself as the good guy, leaving me with only one serious option. I achieved a kind of fame in those early years by my ability, when provoked, to hold my breath for long periods and turn blue. I used to love the stories of twins who, separated at birth, would meet up thirty years later, and turn out to be both wearing double-breasted houndstooth suits and yellow ties with pink spots, both married to a woman called Maria, and both with two children named Buzz and Nancy. I used to imagine that we would turn out to have telepathic superpowers, that, for example, I could drop a hammer on my foot in Cambridge and Unc – at the exact same moment, without quite knowing why – would cry 'Ouch!' in Oxford. Or, at the very least, some kind of unique, special, privileged affinity or synchronicity. But it never worked out like that.

We didn't communicate that well even when we were in the same city. It was a matter of different wavelengths. He was a rocket scientist and I was – well, I wasn't too sure what I was (historian? surfer? poet? failed poet?), but it wasn't rocket scientist. He would carry on surveying the skies for signs of inhabitable planets in far-flung constellations, somewhere out beyond Alpha Centauri and Orion, accumulating vast roomfuls of printouts, while I read *Madame Bovary* and the collected novels of Jules Verne. It wasn't exactly never the twain shall meet, but the twins were definitely twain and their paths intersected rarely. We were like remote planets drifting through space-time. And if I'm ever in pain, the first he knows of it is when I call him from hospital.

We were completely different and yet – unless my American caller was correct and I really was the son of the Queen of Sheba –

shared the same parents. A small mystery that was part of a bigger mystery. I had been idly fantasizing about my own obscure origins and whether or not I was secretly worth a billion, but far stranger and more intriguing was the question of our common origin. How does it come about that sameness can generate difference, or, rewinding the history of the universe, that difference can be seen to fold back into one original source? And, by the same token, was there some natural affinity or continuum between the arts and the sciences (what C. P. Snow called the 'two cultures', well-dressed, well-spoken literary types over here, and white-coated boffins over there), between my brother and me? Perhaps we weren't, after all, doomed to live on separate planets.

In terms of our fraternal genealogy, the answer was relatively straightforward: not being either clones or monozygotic meant that the genetic dice had been tossed all over again, with correspondingly different and indeterminate results. At the same time, I tended to think in terms of a common ancestry for all, no matter how divergent individuals might be in shape, size, colour, temperament. I never ceased to be amazed at the thought, as I sat in a café in Cambridge looking out of the window at all the people passing by in their inexhaustible oddness and diversity, that once upon a time we were all trooping about the same valleys in Africa, or, turning the clock further back in time, floating around the same seas, or, further still, commingled in a star that had itself spun up out of a maternal dust cloud.

Naturally we tried to write a science-fiction novel together, Unc and I. It was a simple plan: you do the science, I'll do the fiction. We had a number of brainstorming meetings either in Cambridge or Oxford. And the novel (called *Soft Life*, RIP) was shaping up promisingly – it told of a software-virus hunter who got caught up in a conspiracy to destroy reality – when it all fell apart. He wanted all the action to take place on the first page and then have 250 pages of technical footnotes, whereas I wanted to anticipate and prolong, postponing the action until the very end. I thought of him as 'my' technical consultant; he thought of me as 'his' punctuation adviser. Perhaps the novel was just too far ahead of its time, perhaps we would have done better to find some form of compromise.

Even so, I remained convinced that once there had been no big difference between literature and science and thought as a whole, that telling stories and the explanations of 'natural philosophy' were inspired by the same set of questions, the same mysteries of our presence on the third planet from the sun. I came to realize that there was one and the same question that all people shared, whatever their different answers, the question of questions. And that was: where did we all come from? How did it all begin? What, in short, was the origin of the universe? Always assuming, of course, that there was a point of origin. It was the Genesis question. A question of universal genealogy, the history of everything and everyone. And I knew that unless you could get the answers straight, nothing else in life would really make any sense and everything would lack foundation. It was the single primal truth that contained all truths, like the catalogue of an infinite library. The Aleph, the microcosm, the key to all mythologies. If only I could lay hold of it, clearly and confidently, I was sure that everything else would be made plain and nothing else would matter. In a word, it all came down to seeing God. That was all.

As Dante put it neatly, at the very opening of Canto 1 of the *Divine Comedy*, which promises sight of God at the end, 'I found myself in the middle of the way, in a dark forest, where the straight path was lost.' I too was stuck somewhere in the middle of things, wandering in 'the dark wood of error'. If I could not know my end with any degree of certitude, I surely could find out more about my – which is to say our – beginning. Thanks to recent scientific advances, and a generation of instruments beyond the X-ray satellites beloved of my twin brother, we are now poised to deliver the final answers about the Genesis moment and solve the mysteries surrounding the origin of everything. To see the light and the whole history of light. And the darkness too. Flaubert said that *Madame Bovary*, which tells of a woman who lived in France some 13.7 billion years after Genesis, was a book 'about nothing'. This too is a book about nothing – and everything.

The twin domes of the Keck Laboratory in infrared light

1

The road sign read: 'BEWARE INVISIBLE COWS'. But I didn't really need it to see that I was in an unusual kind of place.

I was listening to David Bowie's 'Space Oddity' playing on the car radio. 'Here am I floating in a tin can, far above the world'. And something about: 'the stars look very different today'. I felt just like Major Tom. On this day the stars really did look very different. I was finally taking off and leaving Earth behind.

It was like driving from the Garden of Eden to the surface of the moon. As I snaked up from Hilo on the east coast towards the centre of the Big Island, I passed palms swaying in the breeze, giant ferns, banana trees with leaves the size of hammocks, and hibiscus flowers, in yellow and purple, the size of trumpets, sticking their long stamens out at me. A red bird – a brilliant red, all over – flashed by, like something on fire. I drove straight by the Kaumana Caves and the Rainbow Falls. The palatial houses down by the sea and up in the hills shrank down to mere shacks and sheds and finally petered out altogether around mile marker 10. It was raining and everything was a dripping, glistening, glowing green, like it was radioactive. And still I drove relentlessly upwards, up the Saddle Road, into the clouds that hung over the misty mountain. Soon the scenery consisted mostly of infinite variations on the theme of rock. The palm trees had all gone, even the grass had gone. I drove through fields of stone, forests of stone populated by lumbering stone creatures, tribes of stone. I saw flaky rocks, great flat slabs, spiky rocks, frilly ones, fat blobby ones, ones that stuck out at odd angles, rocky cones and cylinders and dodecahedrons, rocks like granola and rocks like chocolate, I saw rocks in the shape of pyramids, and rocks that reminded me of Notre Dame cathedral, and not a few that recalled the hunchback and all the gargoyles. And a few more that looked like plain old boulders, old-fashioned uncomplicated rocks, the size of buildings, in colour somewhere between dark grey and black, tossed down from on high by a careless giant. Here and there huge earth-moving machines were scattered about, boasting massive caterpillar tracks and capacious shovels and giant maws and ten-ton crushers, but they were stationary and lifeless, as if already defeated, feeble and pathetic up against the sheer unremitting, inflexible, and invincible hardness of the terrain.

I am a surfer. Or I used to be. I was in Hawaii, the Himalayas of big waves, but for once I wasn't going surfing. I had been surfing correspondent for a couple of newspapers in England and I had written a book about the feud between a couple of big-wave Hawaiian surfers. Now I was surfed out and written out and I had put away my board and my 'sex wax' and I was getting as far from the beach as it was possible to get. I had spent long enough looking for the elusive perfect wave. I had (as surfers say) paid my dues. I was setting off in search of the truth instead: the original truth, the ultimate truth, the everlasting truth. The mother of all truth. I was here to see God.

Mauna Kea is 14,000 feet high (give or take a few feet), which is roughly half the size of Everest. If you measure it from its real base, on the bed of the Pacific Ocean, it is the tallest mountain in the world by far. Mauna Kea is in fact an extremely large volcano which – fortunately for the inhabitants – ceased erupting a few centuries ago. That's all Hawaii is: a bunch of volcanoes poking up through the Pacific like periscopes. The 'Islands' are nothing but an accumulation of lava flowing forth from a geological wound, the Earth turning itself inside out. The distinctive thing about the Big Island is that it is not only the biggest of the lot but actually getting bigger and possibly higher. And it is high enough, on Mauna Kea, for you to have to stop halfway up in order to acclimatize.

I parked the rental car in the car park of the Visitor Information Station, threw on a coat, and walked out on a lunar landscape. It was a lot colder than at sea level and the air was thinner. What appeared to be large meteorites pitted the lava fields. Maybe I should have been wearing something other than flip-flops and cut-offs. What caught my eye though were all the warning notices pinned up on the wall. This was far enough, they all seemed to say: this far and no further. Please turn back and go down to safer heights. You really don't want to be up here, what with invisible cows and all. Another listed all the 'Winter Hazards' as you approached the summit: wind-chill factors of 40 degrees below zero; snow, ice and impassable roads; violent storms that can last for over a week; people trapped in life-threatening situations. I had already heard about the couple of Japanese visitors who had died on the mountain just a week or two before. I had to stick around at this level for half an hour or more, so I read on.

'Acute mountain sickness is common,' I discovered. Symptoms (according to an 'informational health column') included:

Severe Shortness of Breath
Cough with Frothy or Blood Sputum
Loss of Coordination/Balance
Rapid Breathing at Rest

Vision Impairment
Chest Pain
Slurred Speech
Confusion/Disorientation.

Children under the age of 16, pregnant women, persons in poor physical condition or suffering from heart or respiratory conditions were being strongly advised to turn back NOW! I bought a bottle of sparkling mineral water. There wasn't anything stronger. 'DO NOT DRINK ALCOHOLIC BEVERAGES ON MAUNA KEA'. This was not just a plea for abstemiousness. 'High altitude causes impaired reasoning and drowsiness. Alcohol will further diminish judgment and driving abilities.' The word 'evacuate' came up fairly often, as did 'hazardous'. People found to be suffering from acute high-altitude symptoms had to be transported down to lower altitudes immediately. 'IT MAY BE A MATTER OF LIFE AND DEATH'. It was not exactly 'Give up hope all ye who enter here', but it was close.

I was beginning to think I would need a Sherpa to get to the top. Fortunately, I had one. His name was Doug, he had a generous bristling beard, and he served behind the counter at the Visitor Information Station. He delivered a short sermon on the perils of hypoxia. 'Whatever the altitude, always listen to your body,' he said. 'Don't fight it.' I thought this was good advice.

'My body is telling me to keep going right now,' I said.

'That's good,' he said. He looked out of the window at the cars that were pulling out of the car park and turning back down the hill. 'Most people's bodies get a little cautious at this point.'

Doug introduced me to the wekiu bug (or *Nysius wekiuicola*). He was an unusual little critter who lived on or around the summit and was unique to Mauna Kea ('wekiu' means top or summit in Hawaiian). Built-in antifreeze. Survives by sucking all the juice out of other insects blown up there. 'She inserts the mouthparts into the exoskeleton,' said Doug with a degree of lip-smacking enthusiasm. 'And then sucks.'

'I guessed that's how it would be done,' I said.

It was a cunning evolutionary adaptation, given the lack of edible plants at this height. Apparently visitors were easy prey, knocked out or rendered incapable of resistance by the cold. After all, they were in Hawaii, they were used to warmer climes, they come here for a holiday and then – bam! – they end up as lunch. After a while, I wasn't sure if Doug was talking about other bugs or tourists or, quite specifically, me.

I buttoned up my coat, got back in the car, and drove out of the car park and up the hill. Soon there was nothing but a gravel track winding around the mountain. Down below, fluffy clouds formed a doughnut around the mountain and blotted out the warm Pacific and the coastline. I was looking out for invisible cows: I couldn't see any, and I couldn't hear them either, so they had to be there somewhere, whole herds of them, conceivably, galloping about. As I slid around another bend on the long approach to the summit I spotted another sign stuck in the thick red dust: 'Ice Age Natural Reserve'. It was definitely colder up here but there was a strong sense that I was floating free in time as well as in space. It was like stumbling upon The Land That Time Forgot or going back to One Million Years BC with Raquel Welch (but without Raquel Welch).

As I got closer to the top and the road thinned out and narrowed down into nothing, like a goat-track, and I shoved the gear stick desperately into 'low', I not only had the feeling that it was vertiginous, but also the very word 'vertiginous' kept reasserting itself in my brain, I could virtually see the word 'vertiginous' out there in space, dangling off the edge of the mountain, over the precipice, falling down into the abyss. It was as if the damn word was giving me vertigo. They didn't mention that symptom of high altitude in any of the notices. And there were a few other things they didn't mention either.

The 'Red Square'

2

Intoxication was one of them. I swear I hadn't had a drop to drink. And yet I felt as light-headed as if I had just put away three straight martinis (not, I should add, that I have ever had three straight martinis, but I think I have a pretty good idea what it would be like). It might, in part, have been the effect of swinging around a pillar of rock and seeing first one then two and more spheres, gleaming white or silver, great brilliant domes that you could say mushroomed up out of the red earth on the plateau at the very summit of Mauna Kea except they looked more as if they had landed there, like flying saucers, beaming down from outer space, and still occupying some rarefied dimension not quite of the Earth. They all seemed to be straining back into the void they came out of, arcing around to reach out and touch the sky.

Over here: the James Clerk Maxwell telescope (a 50-foot dish), capable of pinpointing distant dusty galaxies. Over there, the SMA, the 'submillimeter array', the 'Very Long Baseline Array', and the SCUBA: for an astronomer this was the equivalent of having all your Pets of the Month piled up in one neat volume. But I wasn't interested in any passing Pet of the Month, or even year: I wanted all-time, I wanted size, and maximum size at that. To anyone who wants to see the origin of the universe, big is beautiful. Bigger is better and biggest is best. And I was coming to see the two largest optical telescopes in the whole world – and the highest. I pulled up in the extensive shadow of the twin Keck observatories and tried to suck in more oxygen.

Like some kind of novice mountaineer, I had *summited*. My brain was already seriously maladjusted to the sudden leap in elevation my body had just undergone. But as I stood back to admire these two massive, monumental spheres some eight storeys high, I found myself grappling (and struggling, at the same time, not to tumble over the vertiginous edge) with contradictory visions: of a mosque with absurdly huge minarets, pointing the way to God, and then again, intermittently, on another bandwidth of my mental spectrum, of a voluptuous, sensuous sex-goddess, her smouldering, magnificent peaks groaning up into heaven. I was the highest I had ever been. I was utterly dwarfed by everything: minuscule, insignificant, and hallucinating.

Just to be on the safe side, like a nervous examination candidate, I had two pocket-sized notebooks (with pictures of grass-skirted hula dancers on the front cover and gyrating through the blue-inked pages) and three V5 Hi-Tecpoint pens with me. The problem was, these high-tech pens had all exploded on the way up from sea level to the 40 per cent atmosphere up here, spraying purple ink over my bag and, now, down my trouser leg and all over my hands. I think I knew how they felt. My head was spinning too. I was a writer without a pen and I didn't care any more.

Bill Healey gave me a special pen designed not to blow up at 14,000 feet. It had green ink, but it probably wasn't ink at all. A space pen of some sort, either that or it had acclimatized slowly.

'Don't take it down again though,' he said. 'It'll probably implode.'
Bill bore some affinity to his pen just as I did to mine: I couldn't
quite make it up and remain intact but neither could he make it
down again without serious adverse consequences. He told me he
was sixty-something, but he looked around fifty, and a slim,
healthy, and well-preserved fifty at that. He was the kind of guy
who could run a marathon with a 30-kilo pack on his back and then
swim a mile or two just to cool off. But it was as if he could only
really survive in this optimal shape at this level of altitude. He told
me he had been up here for seven years. 'I feel good up here,' he
said. 'You get used to it.' One or two of his colleagues had to wear
small oxygen tanks on their belts and keep taking surreptitious
puffs. Bill Healey didn't. But I had the impression that if you took
him off the mountain and tried to plant him back down at sea level,
he might turn into a pillar of dust. He was like the High Lama of
Shangri-La who could control his own pulse but still needed the
slimmed-down atmosphere and seclusion of these semi-Himalayan
heights to prosper.

Bill told me tales of all the medical emergencies they had had
over the ages, a long-running series of cardiac arrests, cerebral
oedema, something to do with blood gorging on red corpuscles,
coma, and dramatic helicopter rescues. He made me sign a waiver.
I stopped reading after the bit about 'my heirs'. I think it said that
if I died, it was down to me, I couldn't sue anyone. I promised to
forgive everyone. And I would have to pay for the helicopter too.
I thought that was fair: it was my own stupid fault I was up here,
nobody told me to come. It was a vision, a voice, or a calling, but
it was still my vision, my voice, my calling. I had to sit down to fill
in the form, and my powers of concentration and reading skills had
sunk down to barely literate, so he could have been asking me to
sign up for the Foreign Legion and I would still have signed.

It turned out he had been a member of the Foreign Legion, or
the American equivalent, for some 25 years. He still had the buzz

cut. Somewhere in his history he had a degree in forestry too. All of which you might now, in the cold light of day, consider odd and unusual for a man sitting in an observatory on top of a mountain like a fakir up a pole. But at that particular point in time and space, nothing seemed in the least strange or eccentric to me, everything appeared perfectly poised and inevitable, just the way it ought to be and in perfect conformity with the destiny of the universe. When I asked him what his job was, he told me he was 'the anomaly'. Which, again, seemed reasonable to me just then. It was in the natural order of things that he should be an anomaly and that he was showing me around the Keck and that everything had to happen in precisely this way and no other.

Bill didn't search me for weapons (I suspected that he was hoping that I would whip out a gun and then he would have to systematically karate me to death), but he did check to see if I was carrying anything like a bag of crisps, which would likely suffer a messy and explosive fate up here. Crisps were banned, along with ordinary terrestrial pens. At the Keck they seemed to eat only dried seaweed and red iso peanuts. He also offered me a sniff of oxygen, just to keep me going. Like an idiot, a teetotaller disdaining a drink, I turned him down: he wasn't having any so why should I? It seemed to me then (in my ecstatic, cerebrally challenged state) like a form of weakness and self-indulgence. Oxygen? Ha! Who needs it? I'm fine. Never better.

Bill shrugged. 'Mountain climbers call anything above ten thousand feet the "death zone",' he said. He warned me that the eyes are the most 'oxygen-hungry' part of the anatomy, which explained why people were susceptible to hallucinatory states at this height.

'No problem,' I said, drunkenly, starting to slur my words.

He took a good look at me and nodded approvingly. 'You'll probably be OK. You look fit enough. But if you start coughing we're going to get you down.'

I had already marvelled at the latest discoveries of the Keck scientists: a bursting supernova and the incredible and exotic 'Red Square', a highly symmetrical square-shaped nebula (half dying star, half star maternity hospital, prosaically known as 'MWC922') with a clearly defined radiant cross at its centre, like a neon-lit astral advertisement for Christianity or Rosicrucianism. I was, in a sense, following the star in coming all the way up here. But I wanted more: I wanted to get acquainted with the Keck and I wanted to find out how far it could go towards solving the problem of the origin.

The Keck was huge and it was mighty but it was as sensitive and delicately balanced as a ballerina on a tightrope. Any little tremor could throw it off. It was still reeling from the seismic convulsion it had suffered back on 16 October 2006. The earthquake was centred on the Big Island and had shaken up buildings and houses as far away as Oahu. Down below people had been running for higher ground, fearful of tsunamis. Up here there was nowhere left to run. It was the last stop before the moon.

An engineer dressed in some kind of spaceman suit (with 'Craig' on the nameplate) rushed into the control room, took his helmet off, and announced: 'We've got a black-hole leak!' He quietened down and backed off when he realized an outsider was in the room. Nobody else seemed too put out by the news. They took it in their stride. I naturally assumed that the Keck had recently acquired a black hole of some description, perhaps just a small one but dense and mean enough to hold all light hostage, and that they had it neatly wrapped up and under control somewhere, except that now it was leaking. It was all in the order of things.

I knew enough about black holes to know that if it got off the leash and ran amok, it could conceivably swallow the planet and the entire solar system at a gulp. But at this point I had no particular worries and assumed that all this was normal for an average day at the Keck and that they would take care of it and fix the leak before the Apocalypse could kick in.

'Yeah, we got hit pretty good,' said Bill, recalling the earthquake. 'Pointing has been a problem.' Bill was not the only anomaly at the Keck. A good optical telescope is good at doing two things: one is focusing on very distant celestial phenomena; and the other is being able to point very precisely in the right direction. The Keck already had the biggest reflector in the world, 10 metres across, made up of 36 different pieces of hexagonal glass all cunningly woven together like the cells of a honeycomb. But everything else in the laboratory was designed to swivel this very large piece of glass around to target the right little patch of space. And it was the ability to rotate that had been seriously dented by the quake. The vast cylindrical housing of the mirror sits in an apparatus like an extremely large office chair that is capable of swinging around to target the whole of the visible sky. The outer rim, riding on a cushion of encoders, runs along a precise hydraulic 'drive-track'. And it was this groove that had been misaligned by some few elusive millimetres, so that tracking was imperfect. It was like a needle jumping or at least bouncing around in the groove of a vinyl record with a scratch on the surface. Bill called it a 'ding', like a surfer designating damage to his board. He reckoned it at some 45 thousandths of an inch. But it was enough to send the drive-track servos crazy. Any flaw, no matter how slight, when projected out at a distance of hundreds or thousands of light years, was liable to be blown up to the point of obscene deformity. There was no room for imperfection in this equipment. Accuracy, billions of miles away, depended on scrupulous, almost miraculous faultlessness down on Earth. At the Keck, everyone was like monks or saints tirelessly isolating and trying to iron out kinks, sandpapering down the most minor, imperceptible deviations from perfect orthodoxy. There was a purity of purpose about the place that put me in mind of the Sermon on the Mount, Matthew 7:13–14: 'Enter through the narrow gate. For wide is the gate and broad is the road that leads to destruction, and many enter through it. But small is the gate and narrow the road that leads to life, and only a few find it.'

'We've spent months trying to straighten it out again,' said Tomas Krasczinski, who was sitting down in front of a battery of screens, nervously monitoring the patient. He was wearing thick, brown salopettes, he was balding, and his nickname was 'Bulldog', on account of a resemblance to the talk-show jock of *Frasier* fame. Bill, Craig and all the other guys I ran into at the Keck were part of the 'day crew' (some 20 or 30 strong) who constantly upgraded and repaired the gear for the purposes of the 'night crew'. The place was as noisy and dangerous as a building site during the day and only at night turned into a serene institute of higher learning. The bulk of the personnel, in fact, were based down at sea level, in the control room at Waimea, a schoolhouse full of star-struck astronomers, where they spent their time (as I had) inspecting the brilliant images that the Keck transmitted down to them. It was like passing on a flaming baton. But it was up here that the fire was kept burning and the torch could be ignited in the first place. Without the Keck and other instruments like it, the astronomers, the 'remote ops', and behind them the global brigade of cosmologists pondering and analysing and making inferences, and behind them a legion of rank amateur stargazers like me, would have no raw material to work on. They would be like code-breakers with no code to decipher. Bill and his kind, I felt, were the real thing, the serious observers, the sentinels, and everyone else was just hitching a ride, watching the watchers.

'We want more,' said Bill, referring to the level of precision of the Keck, the crucial fractions of a degree that astronomers refer to as 'arcseconds' (there are 60 arcseconds in an 'arcminute', 60 arcminutes in a degree, and of course 360 degrees in a circle, so the arcsecond is 1/1,296,000 of a circle, which nevertheless has to be refined in turn to milliarcseconds or mas). It was obvious that, no matter what, with or without earthquakes, it was never going to be enough to meet his demands, which were potentially infinite. He was like an artist who was never quite satisfied with his last piece of work. Or the insatiable gangster, played by Edward G. Robinson, in *Key Largo*, who confesses to Humphrey Bogart in a moment of self-understanding, 'More. Yeah, that's what I want: *more*.' And he will always get more but, at the same time, will always want more. So he is caught in an endless spiral of desire and frustration.

In a way, Bill Healey summarized the whole history of astronomy right there. If you wanted a one-word summary that would be it: 'more'. From Galileo Galilei onwards, the sufficient *raison d'être* of the telescope and its successors has been: to enable the viewer to see more, and then more again: to go beyond what is visible with the naked eye, and then, going further still, to see beyond (and with greater precision) what the last telescope was capable of seeing. It struck me that the whole point of this artificially enhanced form of perception was to see everything that it is possible to see, to leave absolutely nothing out. The simple criterion that could be applied to the history of cosmology, the theory of everything, was: how far could you go, in time and space? If you could see one mile, why not two? If two, then why not four? There was no particular reason to stop, not until it was simply impossible to go any further, and the end, which was also the beginning, hove into view. This was why I was here: to ask Bill Healey this simple question. How far could he – and the Keck – go?

3

What gives the Keck its edge is not just its technology, which came out of the 1990s (and before that the scientific revolution of the seventeenth century, and before that the Renaissance) and has been constantly refined and advanced through the early years of the new millennium: it is its geography. In the whole of the United States, this was about as close as you could get to the stars, physically, without blasting off into outer space. I suppose you could have tried putting it on top of Everest, but even if other things had been equal, it would still have been a major headache carrying a lens this big up the North Face. You'd have needed a lot of Sherpas. But the other thing about the Keck, which gave it pre-eminence over anywhere on the mainland, was that its atmosphere was far purer. The Big Island of Hawaii suffered less pollution – whether by light or any other human factor – than anywhere else. It had a cleaner, less populated environment. Here, on top of Mauna Kea, we were about as far from the madding crowd as it was possible to get.

The sad truth is that people, and the planets they live on, get

23

in the way of astronomy. Whereas this place was right next door to nothingness and the void. Nothing, or very little, could come between you and the stars. Thus there came about a natural sense up here of intimacy, of affinity, between man – a composite of stardust after all – and stars. If you didn't count that UFO I once saw hovering over my street in Cambridge (and I'll come back to that), Bill Healey was probably about the nearest I'd ever been to a real live extraterrestrial.

'That's pretty cool,' Bill was saying, staring at some image of a far-flung star system. 'But we can get better than that, even finer resolution. It's just an extension of ourselves.' I think Bill was something like the coach of the whole astronomical team at the Keck. He went around whipping people up and stoking their enthusiasm and checking they were doing the right training and making sure they were prepared and peaking at the right time for the next big match.

'Is it going to be ready for Kyle tonight?' he said to one guy.

'I'm ready. Give me two hours,' came the reply.

Bill took me along with him as he went on his rounds. It was the grand tour of this small patch of the universe.

I got completely frozen. I knew I should have put on more layers but the three-martini effect stopped me bothering till it was too late. The air conditioning was tuned to reproduce the same conditions during the day as were predicted for the middle of the night, i.e. bloody cold, on top of a good-sized mountain after all. And there was glycol pumping through pipes beneath our feet, like the opposite of underfloor heating. 'It never gets above freezing in the winter,' said Bill, with hardcore stoical relish, as we made our way along one of the corridors. From his point of view, it wasn't really cold enough. Unless it was freezing it wasn't serious. 'It's the first time we haven't had snow up here in May for a few years,' he tut-tutted. He attributed it to the El Niño effect – it made it a degree or two warmer. But I reckon El Niño must have receded the day I was there because it was as cold as I've been since the time I went surfing in Scotland on New Year's Eve (it wasn't so much the water as the stripping naked on a snow-covered beach). It occurred to me that I could well be the first man to get frostbite in Hawaii. As I looked down at my flip-flops – were the toes already starting to curl up and die? – I found myself envying Bill's rugged boots. I was shivering and folding my arms and stamping my feet to keep warm and starting to turn some shade of blue (and, at the same time, trying not to do any of the above for fear of appearing like a lily-livered beach bum). I realized what it was the great Keck domes reminded me of more than anything else: a couple of igloos, constructed out of blocks of pure ice.

William Myron Keck was a philanthropist. Founder (in 1921) and CEO of Superior Oil, he made a mint out of the oil industry and then, in the classic American tradition, ploughed all the profits into good causes, with a scientific bent. His Keck Foundation now supports (to mention but a few) the Keck School of Medicine, the Keck Computer Science Lab, the Keck Graduate Institute of Applied Life Science, the Keck Center for Interdisciplinary Bioscience Training, the Keck Center for Transgene Research, the Keck Center for Cellular Imaging, the Keck Smart Materials Integration Lab, and so on. It also underwrites *Sesame Street*. Keck is an empire. Twenty years ago the Foundation, now chaired by W. M. Keck's grandson, donated a very large sum of money (in the vicinity of $140 million) to set up the laboratory. Keck 1 began its observations in May 1993 and its twin brother Keck 2 came along three years later. Now Caltech and NASA and a few others own a chunk of it and draw on its results. Which, even though I was more of a stowaway than a shareholder, was just what I was doing.

Bill led me into the dome through a double door: the inner sanctum of the Keck. It was like standing in St Paul's Cathedral, only with a better view of the heavens. Instead of a stained-glass window, K1 was dominated by the huge primary mirror, with its 36 hexagonal segments, poised to capture the faintest furthest photons. The bigger it was the more light it could hoover up, like a barrel collecting raindrops. Ten whole metres in diameter – as big as a terraced house – double (or, for area, quadruple) the mere 5-metre mirror at Mount Palomar (which had been the standard for sheer size in my youth). Opalescent in the yellow light, harnessed in an intricate network of blue scaffolding, like a gorgeous giant butterfly trapped in a web. I was looking up at probably the largest mirror in the world. I was seized by an insane desire to go up to it, pull out a razor, and start shaving, just for the buzz. I had a feeling this was strictly prohibited though, so I didn't mention it.

'The problem with all lenses,' said Bill, 'is it's a big piece of glass. And it's thin. How are you going to hump all that around without breaking it? The sheer weight is liable to shatter it. Which is why we came up with the hexagonals. *Strength*. Every segment was assembled separately and brought up here one by one and then the whole thing was sewn together, like a quilt. Now it functions as one mirror, but we can break it down again if we want to, for maintenance or repair.' Bill rhapsodized about the 'smoothness' of the glass: he reckoned that if you could pull the mirror out to the width of the Earth (and it sounded as if this was his ultimate ambition), the imperfections would still only be a matter of inches high. Each segment is constantly juggled around by sensors and actuators to compensate for the tug of terrestrial gravity and tailored to dovetail with its neighbours to an accuracy of 4 nanometres (about 1/25,000 the thickness of a human hair). The whole thing was held in place by some kind of hairnet contraption that prevented it from either collapsing or floating off into space.

It was about this time that the world started to spin in front of my eyes. Then I realized I was the one who was spinning. I was standing on the rim of the outer shell of the dome, the lens housing. And it was suddenly rotating on silent electric motors, going through its warm-up exercises for the night session. Bill looked up at me with mild apprehension. I was enjoying the ride, but I decided to step off before I was swung round into some immovable mass. 'Good move,' said Bill, with a sigh. 'The tracking sensors are going to be straining to keep the magnets in alignment, even without you hanging on.' Back on the metal deck that runs around the inside, I could see the whole dome rotating on its drive-track. After riding the Keck for a few seconds, I felt as stoked as a surfer after a big wave. But it struck me forcefully that all immobility is an illusion: everything is always moving with respect to everything else, nothing is really stationary, the whole universe is constantly dancing around in time to some inaudible rhythm. Truly, there is no such thing as a body at rest. We felt – as I did, while riding the Keck – that we were pulling off some fine balancing act: the truth was that we were all falling through space at high speed, and getting faster. The difference was that the Keck was falling with style, it was more of a dive with twist.

I don't think anyone at the Keck is actually looking *through* the lens in the traditional way of Galileo and his 'Old Discoverer' or Admiral Nelson putting a spyglass to his eye. A beam of light from the stars is piped around the observatory, like music, bouncing off primary and secondary mirrors, and a tertiary mirror too for good measure, spun around a few electric circuits, channelled through chips, edited, recut, shaped, optimized, orchestrated, and only then recorded on camera and hard disk. Saying, as of some mind-blowing magic act, 'it's all done with mirrors' wouldn't be quite right. You would have to add on what Captain Nemo says about his mysteriously powerful submarine, the *Nautilus*: everything depended on 'a cunning system of levers' – or sensors and actuators, subsystems that were slaves to the mirrors, constantly tending and adjusting and coordinating and caressing. And, in a way, the Keck was a kind of celestial submarine – with Captain Healey at the helm – floating free of landlocked society, hygienically sealed away from the elements, always ready to poke up its periscope and check out the rest of reality.

Bill was particularly proud of the 'adaptive optics', housed inside a huge white box of tricks inside the dome at the back of the primary mirror. The crux of this relatively new addition to the apparatus is a so-called 'deformable lens', rather like the lens of the human eye, that can be refocused more readily than the reflector. It taps into and anticipates all the potential distortions and compensates for them, thus reducing still further the residual atmospheric effects that still plague any instrument on Earth. 'We like to say,' said Bill, 'that it takes the twinkle out of the star.' The other point about 'AO' was that it 'enables us to stay ahead of younger instruments'. I knew what he was talking about, like some old trouper looking nervously over his shoulder at the up-and-coming new generation.

The main force behind adaptive optics was the Hubble Telescope. Named after the great twentieth-century astronomer Edwin Hubble, who first demonstrated that the universe is expanding, the Hubble is a space-based telescope that naturally escapes all the drawbacks of terrestrial telescopes. For the last decade, from an altitude of over 300 miles, it has been sending back to Earth the most dramatic pictures of 'very deep fields' and then 'ultra-deep fields', focusing its mere 2.4-metre-diameter mirror on galaxies more than 10 billion light years distant, peering back in time towards the infancy of the universe. The Keck has had to improve its act to match what the Hubble can do. Rivalry and resentment are a great force for progress in astronomy.

The greatest enemy of the telescope – of all telescopes – is not people or pollution or neon lights: it is the sun. All astronomers are heliophobes, naturally nocturnal. The sun just overwhelms every other light source in the sky. It blinds telescopes as easily as the human eye. While I was inside the inner sanctum, the outer shutters were rigidly closed. 'They peel back when the sun goes down,' said Bill. 'The sun will fry all our electronics. They would literally go up in smoke.' I loved the sheer austerity of this lab. Outside, down on the beach, 14,000 feet down, surf junkies and other deeply bronzed semi-naked people were exposing themselves to the full glare of raw sunlight: up here we were pale anorexics, surviving and thriving on sparse, microscopic scraps – specks – of light, shying away from the jumbo-sized portions of the daily solar binge. It was something like Ramadan at the Keck all the time: we weren't allowed to tuck in until after dark.

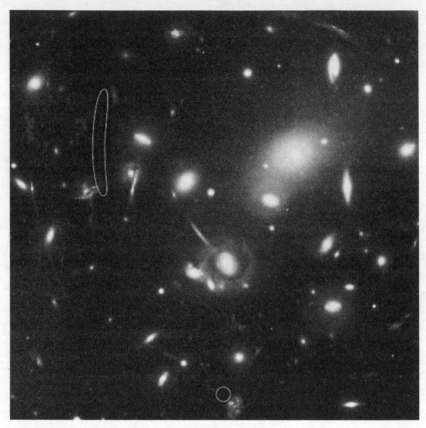

The Red Dot (Abell 2218)

4

Bill took me away and led me through some tunnels. It was a labyrinth down here. I had no idea where I was going but Bill could have done it blind, and presumably, when all the lights were off, did do it blind. Everywhere there were great gleaming racks of tools hanging up, a DIY Aladdin's Cave, enough to fix any conceivable problem. 'We have to be self-sufficient up here,' said Bill. 'Anything breaks, we have to build our own. If we don't have the tool we have to make it.' Down below, in the terrestrial world, everyone chucked everything away, it was a culture of built-in obsolescence; up here, in Keck world, the golden age of fixing – my father's world – in which nothing was ever really thrown away but was constantly patched up, repaired, rewired, resurrected, was still intact.

It was the hexagonal geometry: I couldn't help thinking of the Keck as a giant hive. As we passed through the hidden cells of the structure, I came across drones in overalls fine-tuning odd bits and pieces, soldering them back together, hammering them to an airy thinness, fanatically checking everything over for microscopic

31

smoothness and purity and clarity. Neil was surgically trimming a millimetre off the surface of a hexagonal mirror using a machine the size of a small truck. 'Going up in the big one,' he said, bending down to look along the line of the glass, fixed in a clamp, like a golfer quizzically inspecting the green for irregularities and line, 'when it's done.' He was not yet satisfied. He still had to recoat it in the vacuum chamber with a layer of aluminium two molecules thick. It took me a while to realize that Neil was a technically advanced Keck equivalent of a rank-and-file lens cleaner.

After burrowing down into the basement of the hive we came up again into what would have been the light, if the shutters had been open. I felt as if I was seeing double. This was K2, the second half of the twins, born later with significant improvements, but from the deck it looked identical. The same great glass sitting in its cradle, snoozing, curled up (literally), awaiting the moment when it will wake up and unfurl itself and thrust up into the void. At night K2 would be shooting a powerful red laser beam at a small patch of sodium atoms 60 miles up to provide the telescope with its own artificial guiding star. It was a way of calibrating the cosmos. A lot of people saw it and thought it was some kind of death ray or anti-missile blaster. By this time the cold was biting hard into me, I was shuddering more than shivering as great waves of air-conditioned chill rolled over me. Just at this moment a guy called Mike went by cheerfully pushing the 'ice-wagon', like a cocktail waiter, carrying canisters of glycol coolant or frozen nitrogen to give a quick squirt to anything that was in danger of overheating. They had to restrict 'molecular movement' to a minimum. 'So we freeze the little buggers to death,' Mike cackled, like some Hammer-horror heavy. Some of the electronics needed to be kept at 4 K (4 degrees kelvin), a few degrees above absolute zero. I didn't need to be, though. Make mine a hat, a coat, and a scarf. With extra gloves. I couldn't even take any notes, my hand was shaking too much. I could hardly make sense of what I was seeing any more, it was too far removed from my diurnal sea-level experience. I felt as one does after an entire night spent without sleep: still awake and on my feet, but dreaming deeply.

Bill took pity on me and steered me off to some secret chamber somewhere, with soothing blinking lights on banks of screens, and a much higher temperature. Then he entrusted to me his vision of the future. The Plan. They had figured out a way to finally outgun and outmanoeuvre the Hubble – to turn the pair of telescopes into one even grander more sweeping all-seeing telescope that would blow away everything else on Earth or in heaven. It is what they call the 'interferometer', but it consists of using the two primary mirrors, K1 and K2, as if they were the outer edges of a single much larger lens. They already had the biggest mirror in the world and still they wanted something bigger. It was as if they now had a reflector 80 metres wide, with most of the bit in the middle missing. It was a 'cheat', Bill admitted, but it was a cheat that enabled them to capture more light than ever from ever more distant sources. The two beams met and connected up in the middle of Keck world, in one of the underground tunnels (in fact one of them, K1, had a slightly shorter route, so they kept it bouncing around off a couple of other smaller mirrors for a while until K2 could catch up), and then started comparing notes.

Bill had a dream. He dreamed of hooking up telescopes far and wide, all around the world, thus forming a 'super-interferometer'. The Earth would be turned into a single planet-sized telescope trained on the outer reaches of the universe. An empire of glass. The whole thing would become like an eye, but the eye of God. Then we would have finally attained some kind of omniscience. It was Bill's vision of globalism: one world, collectively focused on outer space, thus transcending our unhealthy self-obsession, our localism, our geocentricity. 'That's all in the future though,' he said, with a trace of regret.

'So how far have you seen exactly?' I said. It was the question

I had been saving up. In a way it was the one question I wanted to ask. It was the real reason I had come here. Everything else was footnotes and polite conversation and dried seaweed. 'What's the furthest thing the Keck can reach out to?'

Bill came to a halt. 'How distant?' He didn't exactly scratch his head, but he did rub his nose. 'I'm going to say . . .' He paused. 'No, hold on. I'm going to have to check on that.' He went over to consult with two or three people sitting at their monitors, going over the setting for that night. They batted it around.

'They all say different things,' he said, finally.

'What's the furthest?' I persisted. I had to know. 'According to them.'

'Come and look at this,' he said.

He sat me down in front of his monitor and tapped a few keys. An image came up of a teeming, tumbling collection of galaxies, a blizzard of light. 'Abell 2218,' he said. 'What do you think?'

'Beautiful,' I said.

'Now look at the ringed circle at the bottom.'

'Got it.'

'See the red dot at the centre?'

'Just about.' I was straining to make it out. It was just a dot.

'Thirteen billion light years,' he said.

I was looking back 13 billion years in time, a mere 750 million years (give or take a few million) after the moment of creation. It was inconceivable, beyond comprehending, but I was looking at it all the same. 'The Keck picked this up?' I hadn't realized that the Keck was capable of seeing that far, going that deep, getting that close to the origin. (I often had to stop and think about the billion part of '13 billion'. When I was a kid a billion used to be a million million. Now it's a mere thousand million. A thousand times a thousand times a thousand. Years. Times thirteen. I realized I could just about get my head around that. I could imagine it. It was around a thousand back to the Battle of Hastings. Another thousand and you were there at the manger for the birth of Christ. You just had to keep on going, multiplying millennia as you went. But now I didn't have to imagine it any more. I was seeing it. All those years and miles wrapped up in the form of light.)

'Well . . .' Bill replied.

The longer answer was that it came out of a collaboration with the Hubble. If you can't beat them, join them. It was a joint Hubble–Keck discovery. The Hubble had come up with the original image, and the Keck had zoomed in on it, in the near infrared range, and given it a bit more definition and substance. In other words, the Keck had hitched a ride on the back of the Hubble. The key technical point was that they had used Abell 2218, a galaxy cluster, to act as a guide star to the more distant galaxy, back around the end of the Dark Ages, and therefore one of the earliest galaxies to be created. It was primeval. The first instalment of photons. It was only a red dot, to my eye, scarcely visible, but for some reason it meant more to me than all the much more spectacular supernovae and red squares put together. In conjunction, the Keck and the Hubble had gone so deep there really wasn't much more light left to observe. They were more or less scraping the bottom of the barrel of the visible universe right there. The image dated from 15 February 2004.

'That's about the limit,' Bill said. 'The most distant.'

'So far,' I said.

'The crown jewels,' he said. 'That's what you've seen here.' He had the same cool, unemotional voice he always had. But you could hear the pride breaking through. 'We're the only ones with the interferometer. We're the biggest – and we're trying to keep it the best.'

'Fuck!' exclaimed Tomas Krasczinski.

''Sup?' said Bill, nervously.

'We zigged when we should have zagged.' He spun round in his chair and looked up at Bill.

'We're out?' said Bill.

'We're out.'

'Shit.'

'I'll get us back in. Don't worry,' Tomas said with unshakeable confidence.

Craig was sitting there in his spaceman suit. I had to ask him. 'How's the black hole?' I said.

'We got it fixed,' he said, imperturbably. It was what he did.

As I was leaving, I bumped into one of the men who had built the Keck back in '90–91. His name was Chris and he was a welder and he was built like a circus strongman. He had a mile of cable lassoed over his shoulder. 'They kept me on,' he said. 'A couple of the tough guys. Anything goes wrong we go in first. Like it's our problem.' He gave a grin and hitched up the cable. 'Which I guess it is.'

5

Star-gazers generally have a good sense of humour about how they are perceived on this largely introspective planet. There was a noticeboard in the common room at the Keck and on it somebody had pinned up a wacky Gary Larson *Far Side* cartoon. It depicted a man wearing glasses with extremely thick lenses, sitting in a café, and gazing with consternation at a spoon that he is holding upside down over his soup bowl. The caption reads: 'Darrell suspected someone had once again slipped him a trick spoon with the concave side reversed.' Maybe it was some kind of allegory about how the engineers up at Keck would find it hard to survive in the everyday world down below. And maybe they would be better off with their bags of dried seaweed and red peanuts.

Bill led me back out into the light of day. It was a shock after the controlled, cloistered environment of the Keck, but I was glad to feel the warmth of the sun on my back again. I could feel my limbs start to relax and unwind. Bill was a fan of Monty Python, it turned out. 'The dead parrot.' He shook his head and chuckled and put an admiring hand on my shoulder as if I had written it. 'That always kills me.'

As I was getting into my car, Bill had a good look at the dedicated four-wheel-drive box. He rammed the stick into 'low'. 'The two Japanese: you know how they died?'

'No.'

'Kept the brakes on for a mile or so and they burned right out. They don't get a chance to recover in this thin air. Car went right over the edge and kept on going. Don't use the brakes.'

'No brakes,' I said.

I drove as carefully as I have ever driven, slithering down the mountain in a mix of first and second, changing gear from apprehension to anxiety attack. This time I decided to ignore the invisible cows. I was looking out for the wekius, but they were so small they might as well have been invisible too. This was where the bug came to chill out. It could only survive on bare rock on top of a mountain. It lived in the cinder cones, it didn't hibernate, and it had become a big environmental cause. No one was supposed to mess with the wekiu. Sometimes the Keck had been forced to back off on building to accommodate the wekiu. That was fair, it was one of the locals. So I didn't want to run one over in the midst of sucking out the insides of its next victim.

For the Hawaiians, Mauna Kea has always been a sacred place, the kind of place where you go to meet God. Some objected to having Keck world up there and said that the mountain had been desecrated. But even with K1 and K2 poised on the summit, it is still sacred. A place of visions. The realm of holiness. Mauna Kea is the point where two beams of light meet and coalesce. As I came down again, descending into the lower realms, I had the distinct feeling that I had just seen God, or Godot, or something wondrous, as if in a dream, but I still had no idea what he or she (or it or they or you) looked like. I had witnessed the sublime and spent all my time shivering as if in a fever. Maybe that was what happened when you witnessed the sublime.

Back at sea level, I headed straight for the Borders Starbucks in Hilo and ordered a double-tall latte and tried to concentrate on the front page of the *Honolulu Advertiser* and failed. As I went around town, back among terrestrials, I couldn't help feeling that the sphere of the sublunary, the sub-Keck world, was a pretty sad place, trivial, contingent, sex-obsessed, conflictual, materialist. I had come down to Earth. It wasn't exactly like dropping out of paradiso and into the inferno, but it was close. I had returned to the realm of fire and tsunamis. There was more oxygen, but I felt heavy and listless, as if gravity had me by the heels and was dragging me down. Like a dead parrot.

Now I was suffering from low-altitude sickness. I was starting to understand Bill Healey.

6

I couldn't sleep that night. I tossed around miserably in my bed at my motel on the outskirts of town. Around 3 a.m. I gave up and took a shower and drove north into the night. An hour or two later, with the first premonition of dawn streaking the sky, I pulled over at a gas station on the coast road, somewhere around Waipio Bay. The young woman behind the counter had just started the day shift. She was all alone. She was the first human being I had spoken more than two words to since coming back down from the mountain. I handed her my credit card.

'Wow!' she said, inspecting it in the dim orange light. 'Is this real?'

'I hope so,' I said.

'We have the same name.'

'Really?'

'Same initial. Except for the "Dr" here it's all the same.'

'What's your name? Your first name, I mean.'

'Angela,' she said.

It was a coincidence, one of those eccentric bonds forged in a far-off land between two strangers. We were almost kindred spirits too. Except that she was an angel (Angela) and I was just a man (Andros). I had a couple of hours to spare and she directed me to her favourite place way out on some cliffs nearby, which looked out over a beach of black sand. She told me that she liked to go out on the edge of the cliffs at night and lie down and look up at the sky and feel the light from the Milky Way raining down on her and hear the waves crashing on the rocks below. 'There's so much I don't understand,' she said.

'You shouldn't have to understand,' I said. 'That is the whole point. There is no understanding. All you have to do is see. That is all.'

I wasn't sure if I was saying this to Angela or to Dr A. Then I went and lay down on the edge of the cliffs, sandwiched between the void above and the abyss below, and listened to the waves dying on the rocks.

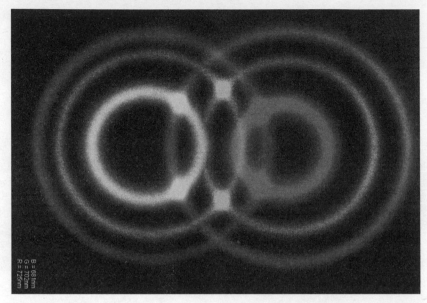

B = 683nm
G = 702nm
R = 725nm

Entangled photons, created by Paul Kwiat and Michael Reck

7

I was trying to telephone a flying saucer. That is how I first got tangled up in all this. It was the beginning of the beginning, the first day.

The real problem with aliens is not whether or not they exist, or whether they are good or evil, but how to contact them across those vast interstellar spaces. (The next problem would be: what are we going to say exactly?) Unc was not really satisfied with the efforts of SETI (the Search for Extra-Terrestrial Intelligence). 'I'm disSETIsfied' was the way he put it, chuckling inwardly, like a repressed Santa Claus. They were way too slow. So far as he could make out, they were relying on standard radio or electromagnetic signals travelling at mere light speed. Just assume, for a moment, that we can fly there (if 'fly' is really the word) or they can fly here. In the end either we or they are going to run into the thorny question of how to transmit a message about what we have seen. It is no good going to another star or another galaxy unless you can send a postcard back home. Imagine that you have just travelled many light years, hundreds or millions of light years. How frustrating is it going to be if you can't crow about it? If you finally discovered extraterrestrial life and nobody knew. By the time any message you sent out actually reached anyone, they would be dead or else you would be. Every message would be an epitaph. Hello/goodbye. So-and-so *was* here. Past tense. I am history, how are you?

Relativity is a major drawback where extremely long-distance communication is concerned. Nothing can go faster than the speed of light (according to Einstein), so if I want to 'phone home', in the immortal words of E.T., and home happens to be Planet X in a system some ten light years away, then it is going to take an irreducible minimum of ten years for my message to get through – and the same for the reply to come back. This kind of constraint is apt to impede transgalactic chit-chat considerably. Fortunately, my twin brother had the answer to this problem. We called it the 'flying-saucer telephone' or (stealing the name from an Iain M. Banks space opera) 'farcom'.

I suspect (I cannot know, since we are not true twins) that it was his underlying sense that there really should be more mutual understanding, harmony and synchronicity in the universe at large, that full-on hardcore twinship should be the rule, that led Unc to build the farcom a few years ago. The objective was simple enough: instantaneity. You had to be able to cut through time, cut it out in fact. Simultaneous messaging. 'Superluminal'. Unc had the theory all worked out. It was a beautiful theory.

'Entangled particles' are the real identical twins of the universe. They offer a bridge across the solitude of infinite space. Entangled (or 'correlated') particles really do behave like telepathic kindred spirits, no matter how far apart they are. They can communicate instantaneously across vast distances. Except, rather like Unc and me, they are antithetical twins: they have a combined spin of zero, which means that if one is up then the other is down (and vice versa). Take two photons (particles of light) emitted from the same source (a laser, say): if particle A down on Earth is jumping through hoops or turning somersaults, then at the very same instant, even though billions of miles away, in another galaxy, on Planet X, particle B will go through the same routine, except in reverse. A mirror image. But A and B dance to the same tune. They have to, they have no choice in the matter, and they do so in perfect unison, like synchronized swimmers. The two particles are so irrevocably fraternal that they will blink on or off, flip up or down, go left or right, with complete simultaneity, wherever they may be. Distance no object. No cause-and-effect delay. There is no cause and effect: the two particles are one, locked together in their passionate embrace across the frozen immensities. Unc didn't exactly say all this, he just wrote down a big, thick, chunky-looking equation, which he thought would say it for him.

$$|\psi(v_1, v_2)\rangle = \frac{1}{\sqrt{2}}\{|x, x\rangle + |y, y\rangle\}$$

Thus quantum mechanics overrides and supersedes relativity. But if this is right, then the lonely occupants of the universe are not, after all, doomed to be isolated monads, marooned, alienated and incommunicado. If you cut out time, then you cut out space too. Entangled particles promised to do a Houdini on those straitjacketed 'inertial frames' (all with different clocks) that Einstein had strapped us up in. Plugged into that snappy interstellar quantum network, the farcom was guaranteed to put us in touch with Alpha Centauri or Betelgeuse or wherever without delay, all those millions and millions of miles away. Distance and separation were an illusion. You are not alone.

That, as I say, was the theory. We called it 'O-theory' (short for 'One' or 'Om'). No one had ever demonstrated it in practice. The time – so it seemed to Unc – was ripe for a rectification. He had done the calculation and come to the conclusion that all previous calculations on the subject had missed 'something' (even though he wasn't too sure what that 'something' was precisely).

It should be clear from the above that if you could only interfere with photon A, make that upspin down, turn it the other way round, then photon B would have to do the exact opposite. Combined spin, zero. And that was all you needed: the binary logic that underlay all possible communication: 1 or 0, up or down, on or off, clockwise or counter-clockwise. To be or not to be. Alpha and omega. Out of that simple opposition arose the most complex communication. Once you had that principle in place you could generate all the possible messages in the universe, all the great works of literature that ever had been or could be written, Homer, Lao-Tzu, Dante, Shakespeare, Flaubert, Borges (and, presumably, the exact opposite, an anti-Homer with a man-eating Cyclops as hero, an inverted Dante with hell where paradise ought to be, a negative Shakespeare, in which the comedies come out as tragedies and vice versa). A couple of photons contained all the libraries in the world, and more. Theoretically.

Unc set up shop in his garage with a view to jemmying open 'a trapdoor in nature'. It was a cryptic phrase that won me over completely. There had to be trapdoors in nature, I was sure. Short cuts, secret passages, wormholes. He knew no one in Oxford or at the Rutherford Laboratories where he worked as a physicist would take it seriously. Communicating with aliens? It had to be a joke, didn't it? In the 90s, the farcom was too far-out, a weird science-fiction fantasy. He was in possession of no grant or government subsidy. Apart from his cat (which, if I'd been making this up, would have been called Schrödinger, but was actually known as Motty), he was on his own. Well, he had me, but I was useless for all practical purposes except as cheerleader and well-disposed observer. I had once fancied myself as a war correspondent, but reporting from the front in Unc's garage would have to do. The great thing about quantum physics is that it insanely upgrades the status of the observer: the presence of the observer actually changes things one way or another, and nothing really happens unless the good old observer makes up his mind to say so. Quantum logic made me feel cheerful about being an observer. While doing precious little beyond observing and making occasional cups of tea, I could finally make a difference. On the other hand, someone has to do the work.

In his spare time, out of his own resources, Unc bought a laser (second-hand), a couple of beam-splitters, and a complicated system of small mirrors. He persuaded our father – for free – to do some precision engineering. Unc took the whole lot and teed it up on a spare tyre (formerly belonging to a Lambretta motor scooter) which was in turn balanced on four breeze-blocks, to maintain maximum equilibrium with respect to the planet. At the CERN labs in Geneva they spend billions on this kind of gear. Unc spent about £100 all told. It was a brilliant and ingenious experiment that would certainly garner him fame and fortune, immense tabloid notoriety, lunch at the White House, a phone call from Steven Spielberg, and possibly even a Nobel Prize, when he got through to some passing extraterrestrial vehicle.

Like a good-luck charm of some sort, Unc had hooked the whole apparatus up to an old telephone he happened to have sitting around: the classic kind of landline telephone, in black, with a receiver sitting on top of a cradle, and a dial with holes for the fingers. It obviously hadn't worked for years. And it looked a little clunky for a farcom. But if it finally rang, then it would be a call from the great beyond. We called it the 'hotline'.

Unc took me over some of the basic presuppositions in the domain. It was like a fast-forward journey through intellectual history. He repeated for me the classic experiment by Thomas Young (dating from 1803) which proved that light is made up of waves. It was simple and yet incredible. Unc took a microscope slide and blackened it up using carbon from a candle. Then he drew a slit in it using a scalpel and a ruler. He shone the laser through the slit: we saw a strip of light, like a narrow window, projected on the wall with a little fuzziness around the edges. The next step was the crucial one. He took hold of the slide again and drew a second slit in it, next to the other one, only about a millimetre to one side. Then he set it up again in its stand and switched the laser on again. Surely there would be two thin windows of light?

I couldn't have been more wrong. The 'interference pattern' that we saw seemed to me magical then, mysterious, almost miraculous, and it still does: multiple, alternating bands of light and darkness, zebra stripes, bright at the centre and gradually fading away to the side. It was the equivalent of dropping two stones into a pond at the same time: the intersecting waves of light cut across one another and either doubled up (crest + crest = light) or cancelled one another out (crest + trough = darkness). Light had to be wavelike because only waves could create interference patterns like these. Young must have thought he had settled the question of the structure of light for all time. And then Einstein came along and demonstrated that light was made up of 'quanta' (Planck's word), little packets of information or energy, which would later become known as 'photons'.

'It was one of his three papers of 1905,' Unc explained (the other two were special relativity and Brownian motion). In this version of things, light was more like a stream of bullets or billiard balls than a wave, capable of bowling out electrons. But no mass, no charge, and an indefinitely long lifetime. 'It was only later,' Unc added, 'that Einstein turned against the quanta. But for the time being he showed that it was a useful way of thinking about light.' Unc had carried out his own experiments on the 'photoelectric effect' to verify Einstein. But Einstein didn't cancel out Young. Now we accept, as Unc put it, the 'wave–particle duality' of light.

Light consisted of tiny chunks or packets (quanta) of information and large, stately, voluptuous curves all at the same time. Light was not 'either/or', it was 'and/and'. Whichever way we chose to speak about light we were invariably leaving something out. There was a kind of built-in inadequacy to our language which light itself did not suffer from. Light, in fact, could not only make even our Olympic sprinters seem slow and lumbering, but it actually makes us look like idiots too.

'The great mystery,' Unc said, 'is how particles know that they are part of some great wave in the first place. How do they know which slits are open?' It was the entangled-particles problem in a nutshell. Unc felt that, ultimately, all particles were entangled one way or another. It was like they were alive. Unc quoted one physicist, Evan H. Walker, who went so far as to argue that photons must be 'conscious, usually non-thinking entities that are responsible for the detailed working of the universe'. Light, in other words, had a mind. Conversely, our minds were made out of the same substance as light, so that when we speak about 'seeing the light' or being 'enlightened' in some way, this is not a metaphor but nothing less than the literal truth.

Unc took the opposite tack: he wasn't sure human beings were really conscious, they were more like a particularly rowdy bunch of particles loosely stuck together. 'The equations don't include consciousness,' he said.

'But don't you need consciousness just to have the equations?' I said.

He pondered that one. 'Maybe you don't even need the equations, come to think of it. The equations don't contain the equations. They're just an add-on.'

'So who is going to call then?' I tapped the farcom equipment to make it clear I wasn't talking about his latest girlfriend.

'Maybe nobody.'

It was a sobering thought. Every now and then, when he

stopped to think about it, Unc admitted the whole farcom scheme was far-fetched and the chances of getting one up and running were limited. He didn't honestly expect E.T. to be dialling his number, not yet anyway. All he wanted to do to start with was a lot more modest (even if improbable). It was the optical equivalent of dropping a hammer on the toe of one particle and waiting for its brother to jump up and down. He took the left-hand photon – we called it Bill – and 50 per cent of the time he stuck a polarizing filter (like the lens from an extremely large pair of shades, which he had in fact taken off a camera) in its path; the rest of the time he let it sail straight through unmolested. Meanwhile the right-hand photon – Bill's twin, Ben – was sieved through an interferometer (a handy barometer of particle behaviour). If everything went according to plan, half the time Ben – reacting from afar to Bill – would make pattern p (polarized) and the other half pattern q (unpolarized). And we, Unc and Andy (well, all right, Unc) would have – hey presto! – bridged the time-space continuum. Lonely planets spinning through space would no longer have to feel quite so lonely. Unc would have proved that, secretly, everything really is connected. If it worked, that is.

8

The great thing about quantum physics is that it makes the theory of relativity seem like child's play. Relativity used to appear mysterious and rather scandalous. It was denounced as 'unintelligible' (in fact even Einstein said at one point that he didn't understand it). But it has gradually become domesticated and familiar. Unc has explained it to me about a hundred times. When I was a teenager you could still find books on the subject ridiculing Einstein and denouncing all his theories. More recently, those books seem to have dried up and disappeared. Hardcore anti-Einsteinians are an endangered species. Now everyone knows that $E = mc^2$ and the speed of light is a constant (well, mostly constant) and a spaceman returning to Earth after a quick flip around the galaxy will find most of his contemporaries in the grave. But quantum physics remains reassuringly exotic and absurd: no one understands quantum physics, not even quantum physicists. Sandu Popescu put me in mind of the words of Tertullian regarding his belief in the risen Christ, *Credo quia absurdum est*: I believe because it is absurd. I met him at the Isaac Newton Institute in Cambridge, a think-tank where the solution to Fermat's Theorem was first unveiled in 1993. He was a big, bright, breezy Rumanian with a chinful of dense stubble, each stalk of which was roughly a trillion times larger than the particles he was so enamoured of. I had gone to see him with an idea of getting a few things straight. Sandu believed.

Einstein, for one, refused to believe. The bizarre laws of quantum physics, letting micro-particles off the leash of macro-laws and inflating the role of the observer, are content to pile up paradoxes. Erwin Schrödinger had devised a thought experiment in which a cat is put in a box together with a decaying nucleus and a 50:50 chance of setting off some toxic gas. In the indeterminate, contradictory quantum universe, Schrödinger's cat is deemed to be neither alive nor dead, but rather some incomprehensible mix of 'superposed states', both alive *and* dead, until the point at which you open the box and take a look (thus, in the jargon, causing 'the wave function to collapse') and the cat becomes either/or. And a tree makes no noise when it falls down, unless there is someone in the forest to hear it. This was the kind of thing that aroused Einstein's wrath. They had to be mathematical hypotheses rather than descriptions of reality. 'Do you really believe that the moon ceases to exist just because you are not observing it?' he fumed. In classical science, repeated experiments must produce identical results; in the quantum lab, results vary unpredictably, particles fly off in whatever direction the mood takes them – or not. For Einstein, this was tantamount to blasphemy: 'God does not play dice,' he famously remarked. Einstein believed that the 'spooky action at a distance' of entangled particles could be explained by what he called 'elements of the physical reality', also known as 'hidden variables' or 'supplementary parameters'. With the 'Einstein–Podolsky–Rosen paradox' (the names of the three co-authors, EPR for short) of 1935, he tried to show that these telepathic twins weren't in fact communicating mysteriously with one another from afar; they were separate, independent, 'local', and simply wired up in advance to be triggered by different interactions. There was no mystery: the whole game was rigged.

'For the next thirty years, physicists turned into philosophers,' Sandu said, shaking his head, as we made ourselves coffee amid the ruck of mathematicians standing around in the common room, comparing notes and chalking up equations on blackboards. And he added, 'Ha!' for good measure. Then, in 1964, the Irish physicist John Bell came up with a new set of experiments that demolished Einstein's supposed hidden variables. He devised a complicated series of hoops, with random permutations built in, for one photon (or rather a small family of photons) to jump through. And still the correlated particle jumped through those same undecidable-in-advance hoops over at the other end of the laboratory. It was the equivalent of Bill guessing what cards Ben was turning over in another room, every time, without cheating. There was no possibility of the twins pre-arranging their stories – they couldn't know what questions they were going to be asked in advance.

The conclusion was inevitable: that entities can interact at a distance as if space and time didn't exist. The phenomenon that came to be known as 'non-locality' – or 'the entangled state' – was real. But then, until the 90s, physics got bogged down in agonizing again over what it all *meant*, the 'what is reality?'-type question. It was precisely the kind of hazy literary-philosophical question that I was rather familiar with. I had often gone about wondering what reality was. I read books about physics like *The Dancing Wu Li Masters*, which depict physicists as Zen masters. Sandu was derisive. 'Don't worry about the cat!' That was his message. Sandu didn't really care whether Schrödinger's cat was alive or dead anyway and ridiculed any questions involving phrases like the 'Copenhagen interpretation' (a classic probabilistic take on the problem). Quantum physicists have tended to be mazy theoreticians, strong on ideas, weak on practicalities. Tantamount to Zen masters, in fact, but with very little talent in the direction of martial arts. Sandu, like

Unc, was one of a new breed, a DIY fanatic of physics, a quantum mechanic who liked to get his hands dirty and wrestle with energetic specks of matter. That was why he was 'the Hewlett-Packard Reader in Quantum Physics' and not just any old reader.

His favourite phrase was, 'As a matter of fact . . .' As far as he was concerned, someone like the cosmologist Stephen Hawking was roughly on a par with a metaphysical poet or a theologian. Any grand unified theory of the universe was just a mirage. Sandu was frankly unconcerned that the whole bundle of quantum ideas was contradictory, inconsistent, incomplete, and bound to collapse like a wave. He was 'relaxed' with it, he said. 'Comfortable', as if he was talking about a pair of old slippers. Mere understanding or the lack of it is a small thing, he thought: why should we expect to understand the intricacies of nature anyway? The more fruitful question, turning Einstein around, is '*Why* does God play dice?' as Sandu neatly put it. 'It is obvious that he does. The interesting thing is – what can we get out of it?'

And his answer was simple: *non-locality*. Non-locality is a gift from God. Non-locality is everywhere, not just in my brother's garage. The beauty of non-locality is that it is the paradox of paradoxes – if you can swallow this one, all the other quantum oddities follow logically. And non-locality opens a whole bunch of trapdoors. Teleportation, for one. Unc had mentioned that if he could get the farcom going, the teleporter would be next on the agenda.

Sandu and I were cycling over to Browns for lunch. It was a mile or so away. Other bikes were flying about unpredictably at high speed, like particles in an accelerator. 'Maybe we should have teleported over?' I said.

'Yes – but I wouldn't like to go first,' Sandu said, sitting down. He explained that the main problem with the teleporter, in its current form, is that 'it destroys the original'. I could see why he would be reluctant to volunteer. It was worse than *The Fly*: at least there part of you comes out at the other end, even if your head is attached to a fly's body and vice versa. 'But the copy is perfect,' he reassured me.

'So it's like a twin?'

'Or a clone.'

You have to think of a teleporter as a three-dimensional fax rather than as ordinary mail: you feed the object in at one end and a replica comes out at the other. So every time Jean-Luc Picard is beamed down to a new planet he dies and a new (but identical with the old) Picard is reincarnated, and yet another beams up again to say, 'Engage – warp 9!' Descartes thought of human beings as machines, but you could just as easily think as them as a bundle of information. Ultimately, if you want to go somewhere, you might be able to text yourself over, or just press 'send'.

All these things were 'at the limit of possibility'. But Sandu told me that he (and a bunch of other guys) had actually carried out an early-days teleportation experiment in a lab in Rome which actually worked. 'So far we have only teleported one photon,' he said. 'Or only the soul of a particle.' The soul had to be re-embodied at the other end. But a principle had been established. They were using entangled particles.

I had the feeling, talking to Sandu, that we were on the verge of a new era. Paradigm shifts in science occur when previously unrelated fields are hooked up together. The nineteenth century fused electricity and light (Maxwell). The twentieth century married gravity and space-time (Einstein). Maybe the big physics of the twenty-first could see the grand synthesis of quantum physics and computer science. The quantum computer would do a completely different kind of maths, not subject to the old binary rules (instead of 1 *or* 0 you would have 1 *and* 0), that can find short cuts and trapdoors through all the old electronic labyrinths and speed up calculations a thousandfold. Computers made out of light. People made out of light.

9

It was funny talking to Unc. When he split up with his wife a decade or two ago, he never really mentioned it. And it took years to get him to talk about it. He was not a great conversationalist (even he would probably accept this description). He had little or no small talk. His school reports often used to complain that he spent too much time 'staring out of the window' and he still had the same distracted air about him. Whenever I asked him what he thought about some recent book or movie, he would counter with a stolid: 'Not bad.' When I asked him what he thought about something I'd written, he'd say: 'Not bad' (sometimes he'd go so far as to add: 'But who do you think is going to read it?'). But when it came to particles and flying saucers and matter transporters, he was a regular fountain of information and ideas. Not quite bubbling, more measured, but consistent. You could hardly stop him. But that day he was content to let the facts speak for themselves.

Unc cast his eye critically over the experiment, like a bespoke tailor sizing up a new suit. Everything was perfectly poised. It was just a matter of mastering the curvature of space-time. That was all. He plugged in the equipment, surveyed the levels one more time, then flipped the switch. Photons went flashing around the mirrors. He leaned down to look through the interferometer. The trick was keeping track of these little balls of light, particularly when they were travelling at 300,000 km per second.

Bill, photon one, was alternately polarized and unpolarized. So Ben – photon two – should be making pattern p and pattern q. The good news is that Ben really did make p and q. The bad news is that he (we couldn't help but say 'he', it's a mistake I know, possibly several mistakes in one, and sexist to boot, but there it is) made them simultaneously, not alternately, so we ended up with a mix, a muddle, of p and q. Instead of being either/or it was and/and. Ben hadn't minded its ps and qs. It was confusing, it was a classic quantum phenomenon. But if you want to report to someone in the constellation of Orion what the weather is like back on Earth, there's no point saying it's sunny *and* raining (even though, come to think of it, that would very likely be true).

The old telephone, the farcom, was not ringing off the hook.

'Looks like there is no trapdoor in nature, after all,' my brother said, more surprised than sad.

'No trapdoor?' For some reason, I had set a lot of store by this farcom idea. It was like a modern Noah's Ark and I wanted to be on it. Now Unc was saying it had just sunk.

'Nope.' He was sounding a philosophical note. But I wasn't feeling philosophical. I was, if anything, desperate.

'You mean you can't talk to flying saucers?' I said. 'No non-local messages? No farcom?'

'No,' said Unc, dashing my hopes.

'Darn it,' I said.

He looked pensive. 'Or rather, yes and no.'

'What?!'

It was sunny and rainy all at once. (This is what it's like talking to Unc.)

The crux is: the particles can send messages to one another all day long, from here to the next galaxy and back, with perfect mutual understanding. But we can't. I had imagined that all we needed to transmit, say, *Finnegans Wake* (since that is made up of an extremely large number of yes/no-type messages) was to impart an up or down orientation to one of those damn particles, to Bill or Ben. But it turns out you can't do any such thing. Bill and Ben are rogue twins, who do what they please. And even if you could impart an orientation to Bill, no one had yet figured out a way to work out what Ben was up to without at the same time changing and corrupting the signal. Every act of viewing betrayed the original. That was the drawback with the non-neutral observer. The message was there but you couldn't peek at it. You wanted to read *Finnegans Wake*, but as soon as you flipped open the book, all the words were immediately scrambled up. (Bad example: *Finnegans Wake* is *already* scrambled up. In fact, take a look at *Finnegans Wake* one of these days: that's how a quantum message looks if you ever dare to open the box. Maybe all books are a little like that.) All you were doing was trying to understand and you ended up producing mindless graffiti. It was as if the 'twins' – the particles themselves – knew what they were up to, but no one else was allowed in on their little secrets.

So even if the phone rang, it was pointless picking up the receiver.

'What would happen if you deliberately wrote garbage to begin with?' I said, getting into the quantum groove. 'Do you think that would work?'

Unc called it – in his cool scientific way (to my immense frustration) – 'non-locality without signalling'. God played dice all right, but not in order to enable lost aliens to phone home. Turns out they were marooned after all. And Sandu's teleportation, similarly, although operating with quantum simultaneity, nevertheless relied on back-up classical messages, the non-instantaneous kind, to disentangle the teleported state. Thus quantum physics and relativity live in peaceful coexistence, not agreeing with one another, but not exactly trying to overthrow one another either.

'I checked that calculation again,' Unc said to me later. 'Turned out I missed out a sign. Happens all the time. A minus sign. When you add it all up it comes out as zero. The two streams of possibility cancel out.'

'So everything you've done adds up to nothing?' I said.

'Like the rest of the universe,' he said, nodding, as if he had never expected anything else. It was a recurrent theme of his, if you could sum the entire contents of the cosmos (counting gravity as negative energy, $- mc^2$) it would come out as zero, everything always added up to nothing. Like the combined spin of a couple of entangled particles. 'I think it might work some day though,' he added, grabbing my attention all over again. 'It just needs something else.' Needless to say, he did not have a clear idea as to what that 'something else' could be. I was left in a band of darkness in which two streams of possibility cancelled each other out, like conflicting waves.

Some time after that day of crushing disappointment, I ended up at a dinner party with both Unc and Sandu and a no-nonsense Portuguese *au pair* called Manuchea (or 'Manu' for short). Sandu had just got back from Geneva. He had been shooting messages 23 kilometres to and fro across Lake Geneva, down a conventional fibre-optic cable, but with a particle attached. The photon was like the secret agent's hair-on-the-door trick: if anyone had tried to intercept the message the photon would show evidence of interference. It would look scrambled. So he was exploiting the very obstacle Unc and I had run into. No one could get away with those old code-breaking Enigma tricks ever again, not without being caught red-handed anyway. But – and I was under no illusions here – these were conventional messages, sent at slower-than-light speeds. *Finnegans Wake*, yes, but it would still take a long long time to reach another galaxy after you pressed *send*. No instantaneity. No trapdoor. We had opened the box and O-theory was dead.

Despite everything, we were still going on about the beauty of entangled particles – they were still sexier than anything else, romantic somehow, in their impossible way – when Manu blurted out the inescapable but embarrassing 'How' question: 'How is it possible for particles to communicate with one another simultaneously across space?' I'd been wondering this myself but hadn't dared to ask.

'Isn't it obvious?' said Sandu. He raised his big bushy eyebrows at the absurd naivety of the question. I think his answer was something like this: for every two entangled particles there must be another couple of invisible particles buzzing about back in time to pass on the good news.

'You're just explaining one mystery with another mystery!' Manu objected. 'It's like trying to explain God and saying it's made up of the Son, the Father, and the Holy Ghost.' She had had a strict Catholic upbringing.

Unc had a different idea altogether. Instead of multiplying the two particles to make four he rolled them up into one. Not one particle but one immense electromagnetic field which is perhaps coterminous with the universe itself. A single all-inclusive giant wave. 'What you have is not two particles, but only one field. The two particles are nothing but quanta – points at which the field is interacting with matter. They don't have to communicate with one another because the field is a totality, always everywhere in touch with itself.'

'You're both mad!' That was Manu's conclusion. 'Physics is stupid.'

I liked Unc's idea of the universe touching itself. Cosmic auto-eroticism. There was a kind of poetry to that. At the same time, I could see her point. But they – Unc and Sandu – had an automatic quantum defence: in a world in which madness is everywhere you'd have to be crazy not to be. Were they not just another part of the universe touching itself? And inside the padded cell of non-locality, there is also the hope of answering one final question, in fact the ultimate question of existence (in Heidegger's resonant formulation): 'Why is there something rather than nothing?'

I had to ask it. There was no way around it. I wasn't sure I understood the question, so there was absolutely no reason why I should understand the answer either. But it didn't seem to matter any more. Matter didn't matter. You could say nothing mattered, but the point was nothing really did matter.

'God created it,' said Manu. 'Or so I believe. I'm just not sure what or who created God.'

'A fluctuation in the vacuum,' said Sandu, 'that got out of hand. A bubble in the foam.'

Unc thought about it for a while. 'If there really is nothing,' he said, 'then there is nothing to prevent there being something.'

His statement was probably coming out back to front and upside down somewhere in a distant constellation at the exact same time and it still made just the same amount of sense.

Sandu said he was planning to make a new universe fairly soon, in a test tube, just as soon as they had created a black hole to mother it.

'So is there any way to verify any of this?' I said. 'Or is it just talk?'

'You don't have to take my word for it,' said Unc. 'You can see the beginning of the universe now.'

'Really?' I said.

'All you need is a powerful enough telescope.'

'You mean you can see God?' said Manu. 'Really?'

10

L ook at yourself in a mirror. What do you see? Something is subtly wrong. You might notice that you are reversed, that your right ear is where your left ear ought to be, your left eye where your right eye ought to be, and so on. Raise your left hand: your image will raise its right hand. Let's say you are wearing a designer suit with the label Armani clearly visible: what you actually see is the word 'inɒmɿA'. This is the inescapable law of left–right reversal, built into the very process of reflection. It is disorienting, it is confusing, but we get used to it. (We anticipate it and write 'ƎƆИA⅃UᗺMA' on the front of an ambulance for the purpose of decoding in a rear-view mirror.) But there is more, something practically impossible to notice and yet fundamental, another way in which our image of ourselves is misleading.

Your reflection is no longer you. You have always already moved on by the time you come to look at yourself, almost as if the mirror has taken a picture of you as you once were. I am exaggerating, up to a point, but the difference between you, as you are now, and you, as the mirror represents you, is measurable.

The reflection is your younger you. This is what you used to look like, a short time ago. And when I say 'short', I mean: very short indeed.

Your smile cannot 'light up a room'. You are not a luminous object. The sun is; so is your bathroom light bulb. You are only illuminated (I am assuming that you have not caught fire recently). A light ray bounces off the tip of your nose. It rebounds off the mirror and is captured by your eye. Even setting aside the time taken by your brain to process the information (remarkably fast), the crucial fact to retain in all this is that light is not infinite in speed. If you could take a very slow-motion film of what is taking place, you could actually see the light ray (or stream or wave) taking off, travelling towards the mirror, hitting it, glancing off, and gradually wending its way back towards you, passing in through the pupil and registering on the retina. Light takes time to go anywhere. Not much time, but enough to make a difference. Usually only small differences, sometimes big differences. In the case of you and your reflection, depending on how far you are standing from the mirror, the difference is a matter of a few nanoseconds (a nanosecond equals one billionth of a second and light travels roughly one foot per nanosecond). But here is the truth: you in the present, you now, as you feel yourself to be, you as pure and unmediated presence, are entirely inaccessible. The you you see in the mirror is the past you, always slightly lagging behind, like a runner who can't quite keep up. The terrible truth is: you are even older than you look. Your reflection is younger than you are. Everything you see is already history. Not very remote history, in fact very recent history, but history nonetheless. That image is a record of a time that has already past. You are no longer you.

By the time you arrive, you have already gone, you have been stood up, by you.

The mirror is a time-machine, but locked in reverse gear. You can go backwards but not forwards. It is a tool of conservation. It shows us things not as they are but as they once were. For example, if you feel yourself dying, hold up a mirror in front of yourself. For a fleeting instant beyond your own death, you will still appear alive, as you used to be, in the glass (even though you may no longer be in a good position to view it). Your existence has passed into the realm of light.

Here is a simple case study. When I finally came down from Keck world and returned to England, I was due to go to see my old friend and neighbour Dr Sidney Abrahams. And when I say old, I mean old: he was 102. We had a lot in common, a love of France, for example, certain books and authors, and we even shared the same barber (he still had a good head of hair). I remember that Marion, his wife, went into hospital for an operation at the exact same time that Heather was giving birth to our first son and that odd coincidence cemented the bonds between us. Whenever in the following years our boys took a tumble and scraped a knee or bumped their heads we would go rushing next door to Sidney for his reassuring diagnosis. It was almost like a laying on of hands – there was something in his calm manner and soothing words and years of experience that children and their anxious parents found instantly therapeutic. And it wasn't just children either. I remember one winter when I had induced major back paralysis by pulling both children along on a toboggan while singing the Russian boatman song – and was duly laid up in bed for a week or two. Sidney and Marion together would drop by each day and would very kindly not laugh at how stupid I had been in the first place. I had many reasons for going to see him but one was that he had been reading a book I had written and I was interested to hear what he had to say about it (he had more to say on these matters than Unc). I was due to go to see him on a Wednesday. On the Tuesday before, Marion phoned to tell me that he had died earlier that morning.

Sidney, after living more than a century, was dead. I could no longer see him alive. No mirror had captured his living self. Only memories and photographs remain. This would seem obvious, self-evident, and yet it is not precisely true. Not true for all time, not true for all space. If there is no absolute time or space, then death cannot be absolute either (although, by the same token, life becomes even more precarious and tenuous and somewhat transparent).

There is a planet orbiting a star fifty light years from Earth and the solar system. Unc would be able to give its exact address, with its cosmological postcode. Its name is (for the sake of argument) Tralfamador. The Tralfamadorians have a very good telescope. They point it at the Earth and zoom in on Cambridge. They see Sidney, aged only fifty-two. He is treating his patients in the hospital and cycling around town and having lunch and going to an art exhibition (clearly the gallery would have to have a glass ceiling for them to observe him observing paintings). On the Wednesday when I was due to see Sidney, if I could only find a way to transport myself instantly across space (perhaps through a wormhole short cut or via a matter-transfer device), I would be able to look through the Tralfamadorian telescope and see him, but rather younger than expected (looking younger, in fact, than I have ever seen him). All I would have to do to see Sidney again, as he used to be, is somehow get ahead of those light rays, speeding away into space, carrying his image with them, like thieves who have run off with an artwork. Sidney lives on, in the form of light, moving much faster than he could ever have done back on Earth. This is the principle of the conservation of information.

To see the past it is only necessary to hold the mirror far enough away from the object. Space is not just proportionate to time, space is time. The further we go through space, or simply see *into* space, the further back in time we travel.

Look up at the moon. (I am assuming it is a clear cloudless night and the moon is full.) The moon is 250,000 miles away from Earth, give or take a few miles. The light from the moon (bouncing *off* the moon, I should say, since the moon is no more luminous than you or I) has taken more than a second to reach you. Astronauts take a few days to get to the moon in a rocket: light makes them seem like they are riding on the back of a tortoise. Light travels at a speed which breaks every speed limit known to human highways: 300,000 kilometres or 186,000 miles per second. (I am rounding up or down the precise figures: 299,792,458 km/s, 186,282,397 miles per second.) It never flags or falters, but it never breaks its own personal best either and just keeps on going at that same steady pace for ever.

Now look at the sun. Better make it the setting sun, to avoid frying your retina. Wait till it is about to slip down over the horizon. Now look at it. You are taking a mental snapshot of what it used to look like some eight minutes ago. You can do the calculation: the sun is 93 million miles away – the distance also known as 1 AU, or astronomical unit – but light travels at 186,000 miles per second; therefore it takes 93 million divided by 186,000, or 500 seconds to go from the sun to Earth = 8 minutes and 20 seconds. (If you build in the other figures which my rounded ones leave out, the exact figure is more like 499 seconds, but let's call it 8 minutes to avoid argument.)

I think I first came across this information in *The Schoolboy's Book of Knowledge*, when I was aged around nine or ten. But it took me years to really get it. If I have got it. But the sun is too close to be really interesting. So let's take another sun, say, Gliese 581c. Gliese 581c is 20 light years away. Astronomers, doing the kind of thing Unc used to do, have recently pinpointed what they reckon is an Earthlike planet orbiting around it. Assuming you could see the Gliese 581c planet, you would see what was happening there not in 2010, say, but rather 1990 (or the equivalent), since light has taken 20 years to go from there to here (one light year is the distance travelled by light in a year, or 300,000 kilometres multiplied by the number of seconds in a year – a long long way). Keep on going, and look at what is going on in the vicinity of a star 100 light years away: you are transported back in time to the beginning of the twentieth century. See that star a thousand light years away? Now you can see what things looked like a millennium ago. Everything is history. The present moment always eludes our perception. You want to go back to the time of the Pharaohs? Check out that star 5,000 light years distant. News of the Pyramids won't have reached them yet.

If I had a telescope that could zoom in on a very large mirror on another planet far away (thus incurring a double time deficit) I could see myself looking a lot lot younger. It would be useless for shaving, but at least I could see what I looked like when I had long hair and a beard, as a student cultivating the Viking look.

The telescope is a tool for inspecting the universe, but in a mood of instant nostalgia.

Are you a student of the Second World War? Would you like to meet Hitler and tune in to some of his speeches? You need to visit a star 70 light years away now. Napoleon? You would need to travel a little further still. All extraterrestrials see us the way we once were just as we see them as they used to be. (Unless they come here or we go there – or we finally get the farcom to work and download Skype.)

Who would like to resolve the debate, once and for all, about the grassy knoll? Who *else* shot President Kennedy? You should all embark immediately (on a very fast ship or using Sandu's teleporter) for a planet 45 light years from here. Historians, I predict, will be the first to sign up for instant travel into distant parts of the universe. Trench warfare, the Depression, the Restoration, the Renaissance, 9/11, the discovery of Tahiti, the first performance of *Hamlet*, Waterloo, the building of the Pyramids, the birth of

Christ, the French Revolution, the Tet Offensive, Henry VIII executing another wife, innumerable wars, massacres, poisonings and assassinations: they are all still happening, in glorious living Technicolor, somewhere out there. They never stopped. Somewhere they haven't even started yet. Somewhere Hitler, Stalin and Pol Pot are just sweet innocent little boys.

We are born, we live, we die: we have a beginning, a middle, and an end. But looked at from a cosmological perspective, not only do we never die, but from certain points of view, we are still not even born. We are born again and again as we pass through new galaxies, and we will die too just as often, only to be born again, and then die. Eternal recurrence is unavoidable. This is not so much reincarnation as us repeating our errors endlessly. We have endless opportunities to start afresh and yet we keep on going down the same path, for ever. We keep on making the same mistakes, for ever. Be careful what you do: your actions will be visible for all time to faraway eyes. Do you seriously think you are going to get away with mugging that old lady and stealing her handbag? Even if terrestrial police fail to apprehend you, whole nations, planets, morally revolted multitudes will be reviling your name for eternity. On the other hand, one good deed and you could be a hero in a galaxy far, far away. Nothing is hidden. There are no secrets. Everything can be seen and known. Everything.

That, at least, is the promise and the curse of the telescope.

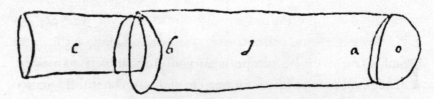

The earliest known illustration of a telescope (in a letter of 1609)

11

The history of the telescope is a history of blindness in need of sight, benightedness in search of light. There are many beginnings. Aztecs, Incas, Inuit, Chinese, Celts, Babylonians, Greeks, the Druids of Stonehenge: many cultures have watched the motions of the celestial bodies, the periodicity of solstices, equinoxes, and eclipses. Knowledge of the heavens entailed power on Earth. But for all the complex monuments to their astronomical concerns – the vast symbolism of the Nazca pampas, temples, precisely aligned slabs of stone, Arab astrolabes, Mayan hieroglyphs – none aimed at artificially extending and enhancing the faculty of vision. Iraqi and Chinese astronomers may have independently developed interesting astronomical tools. But, as the *Cambridge History of Astronomy* puts it, 'a typical Renaissance astronomer looked out on very much the same universe as had his predecessor in Antiquity . . . he had the advantage of reading more, rather than seeing more'.

Jorge Luis Borges, who was blind, wrote that mirrors are abominable because 'they increase the number of men', and he was not alone in fearing their bewitchment. I have been looking into mirrors ever since I can remember (with a mix of curiosity and consternation). I know there must have been a time before mirrors. But even then the precursors of mirrors, accidental mirrors, abounded. We saw ourselves, inevitably, in the eyes of others. Narcissus saw his own face, and fell in love, looking into a pool (in one version of the tale, he is in love with his twin – a girl – and the reflection is a surrogate twin). We caught flashes of ourselves on the polished surfaces of metal objects – shields, for example, the blade of a spear – as we plunged into battle. From a point some thousands of light years distant, we can observe the echo of Phoenician merchants, on a beach on the eastern shores of the Mediterranean, accidentally discovering glass in the grains of sand they use for the purpose of heating food.

According to Pliny the Elder, writing in the first century AD, it happened one evening on the coast of Palestine near the mouth of the Belus River. The merchants set about making supper, but unable to find any rocks to set their pots on, they took some cakes of saltpetre from the ship's cargo and placed their cooking equipment on them. Then they lit a fire. The heat from the fire melted the saltpetre and the sand on the seashore into an unknown fluid. When it cooled and solidified the merchants were astonished to find that it was translucent. So glass was first made. We can assume it was made in many places in a similar way, by chance, but a chance that had to happen, at various times. However it happened, the art of making glass was being disseminated through the Mediterranean by around 3500 BC. Glass was used in place of polished metal to make mirrors. But it was a long time before anyone had the idea that glass could actually improve vision. Saint Paul, in the same era as Pliny, said that 'For now we see through a glass, darkly' (*per speculum in aenigmate*). Glass was synonymous with darkness, obscurity, enigmas. Everybody knew that glass would distort the truth. The best you could aim for was stained glass, decorated with a biblical scene, so that – in some metaphorical sense – you could see God.

No one has perfect sight. We could always do with seeing further and seeing closer and seeing more and seeing more clearly. As with blind prophets like Tiresias, it is our very blindness that generates vision and insight. The defects of glass, it was realized, could be turned around, a virtue made out of a vice. Thus spectacles are born. Some say that Tutankhamun, Saint Jerome and a number of Chinese emperors wore them. Less speculatively, Venice and Florence had become the glass-making centres in the West by the early thirteenth century. Techniques for grinding and polishing glass had attained an advanced state of refinement. A large convex lens was useful as a magnifying glass by older scholars suffering from *presbyopia* (from the Greek *presbyteros*, meaning 'elder'), the condition that affects the ageing eye which is no longer capable of focusing on close objects, such as the letters on this page. Presbyopic readers (I am one of them) find that their arms are too short to hold the book far enough away for focusing. Robert Grosseteste was a pioneer of these 'reading stones'. In *De Iride* (*Concerning the Rainbow*), 1235, he wrote: 'This part of optics, when well understood, shows us how we may make things a very long distance off appear as if placed very close, and large near things appear very small, and how we may make small things placed at a distance appear any size we want, so that it may be possible for us to read the smallest letters at incredible distances.' But the large lens, though helpful, was inconvenient. Glassmakers responded by fashioning small lenses, convex on both sides, that could be fitted into frames. These new glass discs could be seen as metaphorical 'lentils of glass' and hence (from the Latin *lentis*) 'lenses'. Spectacles soon became symbols of learning and scholarship, virtually obligatory accessories for the serious reader and writer.

Nicholas of Cusa (also known as Nikolaus Krebs),

mathematician, astronomer, and learned author of *De docta ignorantia* (which sang the praises of infinite ignorance), may have introduced the use of concave lenses to fix myopia in 1451. But it would take another century before somebody thought to combine convex and concave mirrors and lenses to produce the first telescope. It is possible that an early telescope could have been used to spy out the incoming Spanish armada of 1588 (thus accounting for the legendary coolness of Sir Francis Drake). But the first man shrewd enough to patent the telescope was a Dutchman, in 1608. If I could get to a star 400 light years from here, I would be able to see (with a good enough telescope) the States General of the Netherlands considering the application of Hans Lipperhey of Middelburg and Jacob Metius of Alkmaar to patent a new device for 'seeing faraway things as though nearby'.

The burghers turned down the application. They thought the device was too easy to copy. It was a strange comment since the whole point of a patent application is that you fear that someone is going to copy your idea. Perhaps they suspected that Lipperhey had already stolen the idea from the English. Nevertheless they voted a small award to Metius and commissioned Lipperhey to make several binocular versions of his invention.

The first thing to strike any contemporary viewer of early illustrations of the telescope is the strange resemblance to the erect penis. Inventors of the telescope, it seemed, were trying to extend the range of the eye but on the model of a quite different organ: the telescope was, in effect, an erect, engorged version of the eye. The human eye was good, as far as it went, but it was limited. The telescope made the eye look, on its own, inadequate, misleading, fuzzy. Having eyes was better than being blind, but not by much. Previous philosophers – even the great Aristotle – and would-be cosmologists, it was now clear, had been relying on educated guesswork at best. From this point on, no serious astronomer would think of relying on the power of sight alone. It was a crucial turning point in the history of vision. 'There are more things in heaven and earth,' Hamlet could say at the beginning of the seventeenth century, 'than are dreamt of in your philosophy.' This was not just a pious hope or a dismissive put-down; it was now clear, thanks to the telescope, that there were in fact more things in heaven and earth than had been dreamt of hitherto because it was now possible to *see* them. They had become visible, no longer purely philosophical propositions or a matter of speculation. Seen through a glass clearly, or at any rate more clearly, than before.

The history of science can be simply summarized in two words: 'show' and 'tell'. Unc and I had, between us, more or less set out the state of scientific knowledge at primary school, when we were about ten: he brought in a complicated piece of equipment he had been working on, something like an early anti-gravity device, while I – so I claimed – was exhibiting 'a handful of moonbeams', carefully collected on the way to school, which could not actually be seen because they were so small. (I really was very interested in moonbeams, so it wasn't a complete con.)

Before Galileo, most of astronomy had been confined to telling, weaving together a theory on the basis of shreds and patches of evidence, a vision made up of glimpses and glances. After Galileo, the balance swung decisively towards showing, seeing rather than having a vision. Entirely new things could be seen in the sky, whose very existence had been unguessed. 'If they had seen what we see,' said Galileo of his precursors, 'they would have judged as we judge.' From his performance-enhanced point of view, all previous cosmologists had been cripples trying to run a marathon and in need of a crutch.

While he was in Venice, in the summer of 1609, Galileo Galilei heard rumours from the Low Countries of the use of pieces of curved glass in a tube. He set about making his own, more powerful, version of the new device. In 1610 he published *The Starry Messenger*, or *Sidereal Messenger* (*Sidereus Nuncius*), which proposed unfolding great and very wonderful sights and displaying to the gaze of everyone, but especially philosophers and astronomers, the things that were observed by GALILEO GALILEI, Florentine patrician and public mathematician of the University of Padua, with the help of a spyglass lately devised by him, about the face of the Moon, countless fixed stars, the Milky Way, nebulous stars, but especially about four planets flying around the star of Jupiter at unequal intervals and periods with wonderful swiftness; which unknown by anyone until this day, the first author detected recently and decided to name MEDICEAN STARS.

The name for the moons of Jupiter was a shrewd businesslike move: a tribute to the Medici family, it soon earned Galileo a post as Chief Philosopher and Mathematician to the Duke of Tuscany. Galileo tried out a number of alternative words for the new apparatus too: *perspicillum*, *conspicillum*, *specillum*, *pencillium*. But it was his *telescopium* (*telescopi* in Italian) that caught on (from the Greek, *tele* = at a distance, and *scopein* = to see). Quickly grasping their military potential, he sold the rights (which he did not of course own) for the manufacture of telescopes to the Venetian Senate. By 1611, Galileo was being invited to Rome to show off 'his' invention. By way of aperitif at a banquet given in his honour by Prince Federico Cesi, guests were given a telescope and found they could read an inscription a mile away. For dessert they viewed the moons of Jupiter 'to the infinite amazement of all'. It was the beginning of widespread telescopophilia.

Whether or not it was his invention, Galileo made it his own by virtue of the uses he put it to (and his gift for self-promotion). He had been intrigued but unconvinced by Copernican theory. Now, after further observation of the solar system with a x20 telescope, he swung around behind Kepler and Copernicus and Nicholas of Cusa and asserted that the Earth could not be the centre of the solar system or anything else. We were just one more planet dancing around the sun. 'I hold that the Sun is located at the centre of the revolutions of the heavenly orbs and does not change place, and that the Earth rotates on itself and moves around it.' This was what he wrote in his 'Letter to Madame Christina of Lorraine, Grand Duchess of Tuscany, Concerning the Use of Biblical Quotations in Matters of Science' of 1615. He wrote it in Italian rather than Latin, so that any literate reader could make up her own mind on the subject.

Galileo had opened the box and taken a look. For the first time, he revealed the Milky Way to be made up of myriads of individual stars. In optical terms, he was able to 'resolve' the image, to give a much more precise and defined version; at the same time he was resolving issues and arguments, confirming this theory, blowing away that. The moon was not smooth and perfectly spherical but, on the contrary, he argued, pitted, 'uneven, rough, and covered everywhere with cavities and protuberances'. Even the sun had sunspots. It was radically imperfect, but all the planets paid homage to it anyway. Heliocentrism had supplanted the old geocentric view of things. The sun no longer 'rose' and 'set'. The title of Galileo's 1632 book, *Dialogue on Two Great World Systems*, was misleading or merely polite. The Ptolemaic system and the Copernican system were going head to head, but the argument was already over, so far as Galileo could make out. The Earth couldn't be at the centre of anything. Ptolemy was dead in the water. His proponents were idiots. Galileo was now more than sixty years old. Since he was already under orders not to advocate Copernicanism, it was probably not in his best interests to go into print on the subject. Calling the spokesman of the standard papal position on the subject 'Simplicio' – which was like calling him Simple Simon – was almost certainly not going to garner a lot of sympathy either, not in the precincts of the Holy See.

Pope Urban VIII tried to hold the tide at bay by denouncing

Galileo and putting him on trial, in 1633, for questioning the truth of Scripture and undermining the old way of thinking. It was a classic showdown between showing and telling. The prosecution had a strong case. Had they not already warned him? Had he not agreed (in 1624) to treat Copernicanism purely as a mathematical proposition? According to the Inquisition, Copernicus – and likewise Galileo – by putting the sun at the centre of the solar system, was committing an offence against God: the idea was 'foolish and absurd in philosophy', 'formally heretical' and 'erroneous in faith'. The Bible, after all, was quite specific: 'The sun rises and the sun sets, and it hurries back to where it rises' (Ecclesiastes 1:5). And did not the Psalmist say, '[The Lord] set the earth on its foundation; it can never be moved' (Psalms 104:5)? Would you dare question the word of God? Galileo's defence was that the purpose of theology was 'to teach people how to go to heaven, not how the heavens go'. Also that he thought that his dialogue was really quite balanced and he had never intended to push Copernicanism quite so strongly. This is roughly how the three-way dialogue goes, over several hundred pages:

SALVIATI (very Galileo-like): I think you will agree that, based as they are on strict observations, my conclusions about more or less everything, from planetary motions to terrestrial tides, are necessarily correct and binding.

SIMPLICIO (anti-Galileo): Ha! I prefer to ignore your compelling evidence and stick pedantically to my outdated bookish authorities.

SALGREDO (wise detached arbiter): Gentlemen, I have listened to Salviati's lofty remarks with extreme pleasure and profit and I can't wait to hear more on the subject.

No surprise then that Galileo was found guilty – 'vehemently suspected of heresy' – and his *Dialogue* prohibited. He could have been tortured (he was in fact *shown* the various infernal instruments of torture, the rack, the wheel, the strappado, things with spikes or nails or screws), burned at the stake, or imprisoned; but the judges took a lenient view and placed Galileo under house arrest for the remainder of his life. He was forced to recant on bended knee (muttering 'And yet it moves!' under his breath, according to legend, and certainly thinking it if he didn't say it). When he died in 1642, in Florence, assuming the form of pure light, he was totally blind.

You could pick holes in his arguments. The apparatus was flawed, the curvature of the lens was anything but perfect. You could say it distorted and diminished reality (it did). You could argue that the astronomer was 'bedazzled'. There were anomalies and contradictions and apparent absurdities (out of which Paul Feyerabend builds his anarchistic epistemology, 'Anything goes', in *Against Method*). He had in fact proved nothing. But Galileo *showed* if not the inescapable truth of Copernican theory, then at least that it ought to be possible in principle to show it. *The Starry Messenger* had gone beyond theory. This is what the Catholic Church objected to so strongly: not so much the Copernicanism and the decentrification of Earth as the ability to show rather than merely tell. If you could go this far, putting your eye to a tube and inspecting the moons of Jupiter, then what was to stop you going all the way and seeing God the Father? It was the theological equivalent of being a peeping Tom. Where did God reside? Everywhere, of course, but if it was anywhere in particular it was surely heaven, at the opposite end of the cosmos to hell (with the surface of the Earth poised uncertainly in between, capable of going either way). The telescope threatened to pinpoint his coordinates. The great mystery of the Holy Trinity would be a mystery no more. There would be no more mysteries. And thus no more keepers of mysteries either.

The Inquisition could see all this even if Galileo could not. What did they really care, in the end, whether the sun was or was not at the centre? But seeing God was another story. Had God not said, 'Thou canst not see my face: for there shall no man see me, and live' (Exodus 33:20)? The charge of heresy, in this sense, was entirely legitimate and reasonable: 'correct when based on the facts, the theories and the standards of the time', according to Feyerabend. The Catholic Church certainly took Galileo's propositions seriously enough to sense that they trespassed on sacred terrain. The guilty verdict was the beginning of the rift between the specular and the speculative.

Feyerabend says that the key point is not that Galileo was crazy, but rather that he was 'not crazy enough'. And it is true, he was not going all the way, not even dreaming of going all the way. It was a tentative beginning. On the last page of the *Dialogue*, he concedes that even if he can make sense of 'the constitution of the world', still, on the other hand, 'we are not about to discover how His hands built it (perhaps in order that the exercise of the human mind would not be stopped or destroyed)'. But he had asserted (in the introduction to *The Assayer*) that 'Philosophy is written in this grand book, the universe, which stands continually open to our gaze', even if it was 'written in the language of mathematics' (without which we would be 'wandering about in a dark labyrinth'). Philosophy, knowledge, science, but also theology, they were all 'continually open to our gaze'. The telescope developed in scope and power (with increases in length and magnification and aperture) and went on opening up and refining bigger patches of the night sky. Ultimately, the Galilean shift would open up even the book of Genesis for inspection and revision.

12

So far as I know, Galileo never claimed to have seen God or 'His hands', but he clearly had it in mind. He pointed the way. And, by way of compensation, he did set about calculating the exact size of Satan (he came up with a figure of 2,000 armlengths, and gave a detailed map of the underworld). He was probably the most restless experimenter of his or any age. All conventional wisdom had to be subjected to the rigour of the test. He tested moving bodies by dropping them from high places (possibly not the Leaning Tower of Pisa), he rolled balls down inclined planes, he 'assayed' pumps, pendulums, oars. You might say he was testing the power of the Inquisition when he wrote his heretical book (conclusion: still powerful even if it was the beginning of the end). At some point it struck him that there was one common denominator to all his observations and experiments: light. It was like the underlying premise of all his arguments: without light it would be impossible to see and thus to come to any conclusions about anything, even the Prince of Darkness. It had been staring him in the face all along, like some obvious clue, or he had been staring at it.

If one thing was a given, universally presupposed, it was light. But even light, subjected to Galileo's rigorous questioning, was not devoid of mysteries. To begin with, was it just there or was it in some way in motion? As it came and went with the passage of the days and nights, was it comparable to bodies being thrown off the top of a high building? Would it take a certain time to fall to Earth? Even after his imprisonment, Galileo remained desperate to work out the speed of light. Or at least work out if it had a speed.

He had a suspicion that it couldn't be infinite. He wasn't by any means the first to think this: Empedocles, for example, who ended up throwing himself down into the crater of Mount Etna in the course of an experiment, thought so too. But what Galileo lacked was evidence to support his feeling. It was impossible to *see* the speed of light. But if he was right then surely it would take a measurable amount of time for it to travel a mile. So, in Tuscany in 1638, when he was almost blind, more obsessed with light the less he saw of it, he set two men a mile apart from one another, standing on hilltops, each equipped with lanterns, and each lantern equipped with shutters. When one man opened the shutters of his lantern, the idea was that the other would open his as fast as possible. Then the first observer would measure the time difference between him opening his shutter and seeing the second light.

The experiment was one of Galileo's most outstanding failures. You could blame the human factor: all the reflex times were greater than any possible differences due to the speed of light. All you could measure, at best, was how fast observers reacted. After a lot of practice, they got it down to a pretty fine art. They tried standing further apart. But light just rolled on regardless, unstoppable, imperturbable, seemingly immeasurable. The best Galileo could offer was that light must be at least ten times faster than the speed of sound (the same observers could make a noise at a distance of a mile and that definitely took a lot longer to reach anyone, irrespective of reaction times). But all that meant was that anything much below 8,000 miles per hour was excluded, but that everything else was possible, even, logically, an infinite speed. Kepler for one thought light was instantaneous, crossing the universe in a flash. It was ironic that another of Galileo's failed experiments – in a sense all of his most successful ones too – already contained the answer he was seeking.

Galileo loved to watch the Venetian ships come and go in the harbour at Padua. He worked out the physics of rowing, correctly adducing water itself as the fulcrum of the oar. He was no navigator himself, but he ended up teaching navigators how to steer themselves around the globe. Which is how he came to bump up against the conundrum of longitude. For the seventeenth century, the era of the 'scientific revolution', it was a major embarrassment. Latitude (the parallel lines on a grid, north and south of the equator, like horizontal hoops) had been worked out by the Portuguese, and required only an astrolabe and a cross-staff (forerunner of the sextant) to get a fix on the sun and the stars and work out location to within one degree; but longitude – marking a shift in time as well as space (one hour equals 15 degrees, east to west) – had defeated the finest minds on the planet. Nobody was quite sure where America was, nor where China and Japan were, nor even how wide the Mediterranean was. It was clear that a solution would greatly improve cartography and simplify navigation (and thus bring greater glory and possessions to any one of a number of rival empires). Philip III of Spain had offered a prize – a life pension – for the first method to determine the longitude of a ship out of sight of land. In 1613 Galileo put in a bid for the prize. His solution hinged on the eclipses of the moons of Jupiter, the 'Medicean stars' as he called them in *The Starry Messenger*.

Galileo had been the first to discover that Jupiter had four satellites, orbiting around it as planets orbited the sun. Eclipses of the moons occurred regularly and frequently, one thousand times annually, as the moons passed into the shadow cast by the mother planet; therefore, Galileo reasoned, it should be possible to use them as a kind of cosmic clock. After months of patient study, he came up with a set of tables that predicted each of the moons' appearances and disappearances (which he called 'ephemerides'). Thus a comparison between the local time of the eclipses and the standard schedule would indicate the longitude of the navigator. Galileo even invented a special helmet with mini-telescopes attached – a 'celatone' – dedicated to observing Jupiter and its moons. A brilliant device whose only drawback (as Dava Sobel points out in *Longitude*) was that the mere pounding of your heart could cause the whole of Jupiter to bounce right out of the field of view.

Galileo didn't win the prize (although the Dutch – the same States General that had denied Lipperhey his patent – later awarded him a gold medal and chain). Perhaps Philip III thought the idea was too exotic, too impractical: could sailors really become astronomers? Wouldn't clouds get in the way? Still no one improved on the method until the middle of the eighteenth century and the development of the chronometer and the work of the English clockmaker John Harrison, thoroughly trialled by Captain Cook. The English had improved on the original telescopes, and had gone as far as installing gimballed observation platforms, immune to the motion of the ship. The fact was that there were far too many inaccuracies in the timetables for them to be of much use. Even Galileo himself admitted that there were apparent inconsistencies in his data that he could not seem to iron out.

But it was out of these inescapable errors that the truth about light would ultimately crystallize and shine forth.

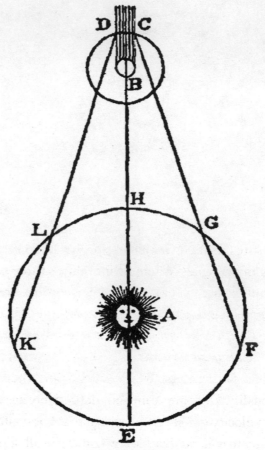

Fig. 70.

Rømer's view of Earth and Jupiter and Io

13

René Descartes trashed Galileo in a review of 1638: 'his fashion of writing in dialogues, where he introduces three persons who do nothing but exalt each of his inventions in turn, greatly assists in [over]pricing his merchandise'. The author of *The Discourse on the Method of Rightly Conducting the Reason and Seeking Truth in the Sciences* and the *Meditations* was Europe's chief natural philosopher of the age. His phrase *Cogito, ergo sum*, 'I think, therefore I am', defined rationality. Descartes came out definitively against Galileo. The infinite velocity of light 'is so certain that if it could be proved false I am ready to confess that I know nothing in all of philosophy'. Light was close to God and therefore could not be bounded. You might as well try to subject God to experiment as try to test the speed of light. But not everyone had given up on Galileo's dream of capturing light and running a slide rule by it.

Perhaps Rømer was still dreaming of the elusive longitude prize. But the thing that perplexed and obsessed him over many years was how it was possible for Galileo (and his successors) to get the ephemerides so wrong, so systematically wrong.

Ole Christensen Rømer was born in Aarhus in Denmark in 1644, two years after the death of Galileo, and studied in Copenhagen. He edited Tycho Brahe's manuscripts and was appointed a court astronomer in France, teaching the Dauphin, and travelling around France making observations for the Académie. He became professor of astronomy and mathematics back in Denmark but he remained an immensely practical man: he was master of the mint, a surveyor of harbours, an inspector of naval architecture, head of a highway commission, and expert on ballistics. He didn't always have his head in the clouds. He liked things to be precise and predictable. He built clocks and thermometers as well as observational instruments (and ultimately, in 1704, his own observatory). He abhorred lawbreakers, and became tax assessor, magistrate, mayor and senator. He was convinced that there were laws that governed the universe, just as laws also governed the realm of society. The planets and the stars had to obey the law in just the same way as the King's subjects. There could be no exceptions. And, by the same token, everything should be measurable. With this in view Rømer invented a new system for weights and measures and devised a temperature scale before Fahrenheit and revised the Danish calendar, so that Easter was scheduled in accordance with the rhythms of the moon.

Which explains why Ole Christensen Rømer found the business of the moons of Jupiter so irritating. Io seemed, on the face of it, to be deviant. A lawbreaker.

In 1671 Rømer joined the observatory of Uraniborg, Tycho Brahe's 'heavenly castle' on the island of Hven, near Copenhagen. Over a period of several months, he took notes on no less than 140 eclipses of Io. At the same time, Giovanni Domenico (or Jean Dominique) Cassini was observing exactly the same eclipses at an observatory in Paris. They were both equipped with the latest telescopes, some 30 or 40 metres long, with crosshairs on the focal plane, and extremely precise clocks, trustworthy to within a quarter of a second. All they had to do was compare the times of the eclipses to work out the longitude between Paris and Uraniborg. In theory.

In 1672 Rømer was invited to Paris to join Cassini as his assistant. Cassini was the director of the new observatory, supported by Louis XIV (the 'Sun King'), and a world authority on Jupiter, having discovered its Great Red Spot. He had also discovered a couple of the moons of Saturn and would have part of its rings (and, ultimately, even a spacecraft) named after him. He would produce the seventeenth century's most detailed map of our moon. Testing Galileo's theory of longitude had been his initiative. And the new timetables – the revised ephemerides – were his. Rømer was only 27 years old, the equivalent of a young postgrad, but bursting with self-confidence. Cassini was widely respected, covered in honours, and well-off; Rømer was an unknown and impoverished.

All the same, Cassini had to admit to his apprentice that there were still discrepancies in his measurements. For the next five years, Rømer dedicated himself to ironing out the discrepancies. He noticed that over several months, the predictions became more and more inaccurate, like one of the cannon that he studied that had swung off its bearing, till they were off by some eight minutes. And then, just when he was starting to give up on the original model, he found that it was becoming more accurate all over again, to the point that theory and observations coincided perfectly once more. So the original calculations had been correct after all. But would they stay correct or go haywire once more? There had to be some explanation of what had gone wrong. The telescope itself? He checked it again and again. There was nothing wrong with it. Could it be the moon itself, jumping around in its orbit? Ridiculous – moons did not jump, they described perfect ellipses, mathematically measurable.

There were errors, but there was a pattern, a regularity, to the errors. The predictions were wrong but in a way that was itself predictable. The difference, Rømer came to realize, could only be accounted for by Rømer himself, in his status as observer. Rømer was solidly planted in Paris. But the Earth itself was orbiting around the sun (in this he followed Copernicus rather than Brahe). As it swung around the sun it would be closer and then further away from Jupiter (also orbiting, but much more slowly). The two planets together performed a kind of dance, now close, now further apart. Rømer himself was part of the dance. The greatest distance would be when they were on opposite sides of the sun, the shortest would be when they were close together, on the same side. And light had to travel across space from the moons of Jupiter to reach Rømer in France. But if the speed of light could not be measured on Earth itself – as Galileo had reluctantly shown – perhaps it could be measured across the vastness of interplanetary space. Cassini, on the other hand, sided with Descartes and firmly rejected any notion of light having finite speed.

In 1676 Rømer presented his findings to the French Academy of Sciences under the heading 'Démonstration touchant le mouvement de la lumière trouvé par M. Rømer de l'Académie des sciences' (published 7 December in the *Journal des Sçavans*). The young Dane completely demolished the conventional wisdom. Rømer compared the view of the eclipse as the Earth moves towards Jupiter (F to G in the diagram) with the view as it moves away again (L to K). The eclipses occurred ahead of time when the Earth was nearer to Jupiter, and fell behind time when it was further away, with a total difference of 22 minutes over 40 orbits. The only explanation, Rømer asserted, was the time taken by light to travel all the way from Jupiter to the Earth.

With a tremendous flourish, in the following September, Rømer was able to predict to the Academy exactly when Io would complete its next eclipse. On the evening of 9 November 1676, at 5.35 and 45 seconds, the moon emerged from shadow, from the point of view of the observatory in Paris. Precisely as Rømer had predicted; and 10 minutes later than predicted by Cassini's old tables. The mystery of the unpunctual eclipses had been solved. The old school of Cassini and Galileo, the men with beards, had been overthrown by the young whippersnapper. The members of the Academy were convinced by the demonstration. The excitement of the reporter who wrote up the story in the Academy's journal was easy to understand: it was as if the 'emersion' into light that Rømer spoke of was being enacted, intellectually speaking, right before their eyes. The mysteries of Io had been brought down to Earth. All obscurity and darkness had been banished by Rømer's explanation of 'the retardment of light'.

It should have been easy at that point to work out the speed of light. It would have been the first universal value to be established on Earth. The calculation was eminently doable (speed = distance over time). The odd thing is that Rømer himself never in fact undertook it or never published it. Having shown it must be possible, he then shied away from the result. Christiaan Huygens (who wrote the *Treatise on Light* of 1690) was the first to do it, the following year, and even he got it wrong on account of misunderstanding what Rømer had said to him. But perhaps Rømer wanted to be misunderstood. It was almost as if, having made the discovery, Rømer wanted to cover it up again, as if it was something almost to be ashamed of. Perhaps it was as Fontenelle had suggested, in his history of the Académie, that 'If we deduce from [the observations of M. Rømer] the distance that light travels in a minute, and the distance it must traverse in order to cause a ten-minute delay in the phenomenon, we shall be terrified both by the immensity of the distances, by the rapidity of the movement of light, and by the disproportion between this speed and that of human industry.' Perhaps Rømer felt like the philosopher Pascal, who wrote at the same period: 'The eternal silence of these infinite spaces frightens me.'

There was a swift rearguard reaction against Rømer. Somehow Cassini managed to maintain that the Dane had neither proved nor disproved anything. In England, Robert Hooke dismissed Rømer's light: he continued to maintain that "'tis so exceeding swift that 'tis unimaginable' (1680). Newton, in contrast, appears to refer to Rømer in his *Principia* (1687): 'For it is now certain from the phenomena of Jupiter's satellites, confirmed by the observations of different astronomers, that light is propagated in succession and requires about seven or eight minutes to travel from the sun to the earth.' But this was like a late addition and Newton hadn't fully absorbed the implications into his theories. For all practical purposes there was no speed of light. In the Newtonian universe, everything worked like clockwork, all the parts fitted together into one neat, tight, well-assembled box. Gravitational effects, of the Earth on the moon and the moon on the Earth, were instantaneous. Each part could see what all the other parts were getting up to instantaneously. Any observer, anywhere in the universe, could see what everyone else was doing, simultaneously, at exactly the same time they were doing it. There was no slippage, no delay, no gaps. Everything was smooth and well-oiled by the free passage of light. There were no dark corners, no nooks and crannies of existence. God was omniscient and there was no reason why, in just the same way, an observer (slightly idealized) could not know everything that was going on. Ignorance was just a temporary failure of the human intellect, there was nothing inevitable about it. God had said, Let there be light; and there was light; and that light was everywhere, for ever, instantly. What he did not say was, Let there be light but it might take a while to reach you, depending on where you are precisely.

Edmund Halley upped Huygens's estimate of 'the equation of light' after determining that Rømer's 22 minutes should be more like 17. But, despite everything, the theory that light had a finite velocity was not fully accepted or validated until well after Rømer and Newton, in the work of the British Astronomer Royal James Bradley in 1727, with what he called 'the Aberration of Light'. And it wasn't until 1809 that Jean Baptiste Joseph Delambre, still drawing on observations of Io, worked out the speed as being in the vicinity of 300,000 kilometres per second. Further refinements followed with rapidly spinning toothed wheels and rotating mirrors.

The 'Demonstration' was a brilliant insight by an astronomer, who was able to see the truth and not theorize his way around it. But having seen it, Rømer chose to ignore it, took no real notice of it, and returned to Denmark (not surprisingly Cassini didn't want him in Paris any more) and went back to more immediate priorities, like measuring the size of roads and paving them and propping up flood defences and fixing the 'Danish Mile' at 24,000 Danish feet, based on astronomical constants. The rest of his life was dedicated to propping things up, as if he suspected that the speed of light would throw everything off, no matter how slightly, and he was personally obliged to stave off disaster.

Rømer's observations were lost in the great fire of Copenhagen in 1728. So we can only speculate, but it seems certain that he must have realized that in looking up at the night sky we are not seeing things as they are, but only as they were, and have been, and always will be; and that it makes a big difference where you are located in the universe. A point in space was necessarily a point in time too. It was the longitude problem writ large, writ very large, on the scale of the cosmos. A degree of uncertainty had entered into the astronomer's mind. Pascal had described the universe as 'a sphere whose centre is everywhere and whose circumference is nowhere'. After Copernicus, the centre had been lost, it was everywhere and nowhere. But after Rømer, perhaps, in some obscure way, it was capable of being restored – not in space but in time: back at the beginning, the singularity from which everything originally emerged, and that focal point on which all gazes converged.

In 1705 Rømer was made Chief of Police and until his death in 1710 he made it his life's work to reform and control beggars, prostitutes, and tramps. Conscious though he was that there would be an inevitable (if very slight) delay in the transmission of light, he installed oil lamps on the streets of Copenhagen to banish the darkness from the Earth.

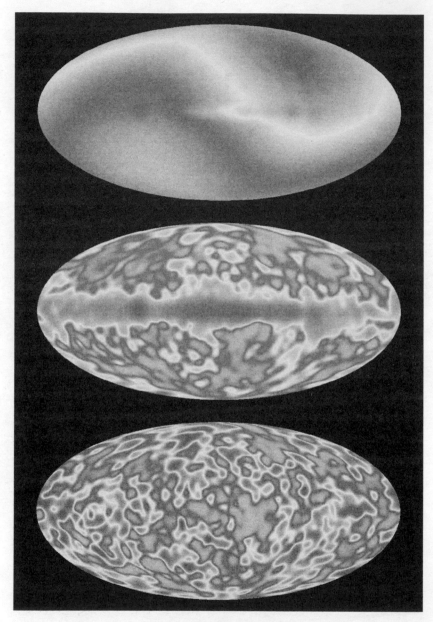

The 'face of God'?

14

'Where are you going, Uncle Andy?'
    'I'm going to see God.'
'You know you can't see God.'
'Well, I'm going anyway.'
'Can I come?'
Which is how my Australian niece Juliet happened to come with me to the Cambridge Astronomy Laboratory. It was just a bike ride away. She was 17, on a world tour, and loved cats, epic tales, and chemistry, biology, and physics. And she was a real tiger for mathematics. Once, when we were having lunch together in a restaurant, and one table began singing 'Happy Birthday', I wondered out loud how many people in the restaurant would be having a birthday on that day. She covered two napkins with complex equations working it out and insisted on doing it the hard way with no short cuts (her final answer – if it was final – was something like:

$$P(X \geq 2) = 1 - (364/365)^{10} - (363/365)^9(1/365) - (364/365)^8(1/365)^2).$$

Telescopes had come a long way since the time of Galileo. For a century or more Cambridge had been at the forefront of a lot of the developments. Less than a mile from King's College Chapel and Newton's old rooms by the Great Gate in Trinity, the old observatory stands at the end of a tree-lined avenue off the Madingley Road, tucked away among leafy courts and lawns and ivy and vast spreading oaks, perched on a hill with a classic dome-shaped housing. A massive cylindrical lens periscoping up into the heavens. It harks back to a golden age of chilly nocturnal star-gazing when looking at the sky was like communing with God and finding peace in your soul. I remember seeing Io for the first time here: Io filled the lens and my consciousness and vacuumed out of my mind everything that was not one of the moons of Jupiter. There was no more 'I', only Io.

It was built like a temple. The Duke of Northumberland made a gift of the first telescope in 1823: an achromatic doublet object glass of 11.6 inches, clear aperture, focal length 19 feet 6 inches, made by Cauchoix of Paris, kept in perfect tension and protected from torsion and flexure by two massive triangular prisms designed by Troughton and Simms of London. The polar axis frame and the main casing were made out of specially imported Norwegian fir. For the best part of a century it was the biggest refracting telescope in the world, driven by a fanatically precise clock-driven equatorial mounting capable of tracking a star in its diurnal motion across the sky. The perfect instrument for scrutinizing the inverse-square Newtonian universe, that neat clockwork box with stars rotating inside it like cogwheels, which needed a divine clockmaker to make it tick and set it spinning in its precise orbits. Newton's cosmos was a theatre in which all the dramatis personae had their parts carefully scripted in advance and nobody ever deviated or ad libbed or fluffed their lines. And the Cambridge telescope gave a view from the royal box (so it was that Cambridge's astronomer would be known as the Astronomer Royal).

Even if the Cauchoix was a collector's item, harking back to a more aristocratic age, an extremely large museum piece, it was in fact still perfectly functional (an electric motor, replacing the original clock, still turned the polar axis once every sidereal day). It was a fine and freezing winter's night and Juliet and I watched a meteor flaring up and then dying out overhead before putting our eyes to the lens and admiring the rings of Saturn. Then we stumbled in the darkness over to the Thorrowgood, probably the largest (traditional) telescope in the country. We had a good view of several billion stars in the Andromeda galaxy.

'What do you think?' I said.

'Well, I'm not seeing God yet,' she said. We had recently been to the Old Bailey together and watched a murder trial taking place and she was much preoccupied with the concept of evidence.

'So,' I said, determined not to be completely outmanoeuvred by a 17-year-old girl, 'you know what God looks like then, do you?'

The Cambridge telescopes had been intrepid explorers of the night sky. They had filled in a lot of gaps in the astronomical map. That night, there was a big, bearded man called Olaf giving a talk about the size of the universe. He was standing on top of the hill looking up at the stars. I asked him where the biggest telescope on the planet was, the one that would give me the best view of everything. 'The Keck,' he replied without hesitation and with a note of respect. 'Hawaii.' But he added that if I could only hang on a few years, there was a new even bigger telescope due to come online, with a lens double or triple the Keck's, some 20 or 30 metres across. It was going to be called 'The Extremely Large Telescope'. Further down the road there was talk of telescopes 50 or 100 metres wide. 'What are you going to call those ones?' I said. 'Ultra-Large Telescope,' came the unhesitating answer.

But there were other ways to go, even within Cambridge. The Mullard Radio Astronomy Observatory, a few miles further out of town, made the old optical telescopes look old. No Greek temples, no domes, no precision-engineered clocks. The new generation of telescopes, after the Second World War, looked more like a cluster of electricity pylons with a large bowl perched on top. They were erected on the bed of the disused east–west Cambridge–Bedford railway line. And they kept on expanding from the Small Array to the Large Array to the (inevitably entitled) Very Large Array, hooked up to a new generation of giant computers called 'Titan' (for running inverse Fourier transforms). The 'One-Mile telescope' gave way to the Ryle telescope (named after Gilbert Ryle, the founder of radio astronomy), also known as the '5-kilometre telescope': a network of vast dishes mushrooming up out of the Cambridgeshire countryside, some mounted on rail tracks for mobility, all tilted up towards extremely remote nebulae and gas clouds, swivelling to track the complex paths of binaries and register elusive Doppler effects, like a giant magnifying lens peering back through time and space at far-flung but deeply significant footprints among the stars.

Radio astronomy opened up another perspective on the heavens. All those secret signals from mighty stars a million parsecs away. Electromagnetic manna pouring down out of the great luminous sky. And all for free. It was like winning the jackpot every night. All the giant antennae had to do was gather in the infinite data streams, like rainwater in a barrel. I always thought of them as beautiful, the way they thrust up out of the land into the broad air and swung around like sunflowers, turning towards the light from suns not our own, endlessly ripping, carving, surfing the microwave spectrum. Radio telescopes in Cambridge had notched up some notable achievements over the ages: in recent decades, for example, the discovery of pulsars – the signals from neutron stars, the fast-spinning remnants of collapsed supernovae, so dense that a teaspoonful weighs as much as a small planet – in 1967, using almost amateurish equipment, a few acres of poles strung with wire and dipoles. This was mainly down to Jocelyn Bell.

She was only a twenty-something grad student at New Hall. She had joined in hammering a thousand poles into the fields, as if she was putting up a circus big top, and stringing 120 miles of wire between them. She was looking for fluctuations in radio emissions ('interplanetary scintillations'). She was the one who scanned the printouts, from four three-track pen recorders, 96 feet every day. She was the one who started to see what she at first called 'unclassifiable scruff'. She soon realized that this was a regular signal, like a pulse, being emitted from 'right ascension 1919' every 1.2 seconds. Her first thought was: flying-saucer telephone. She was worried: she didn't think it would look good in her doctoral thesis that she thought she had tuned in to an alien civilization. 'I went home that evening very cross,' she said later, 'here was I trying to get a Ph.D. out of a new technique, and some silly lot of little green men had to choose my aerial and my frequency to communicate with us.' Which is why they originally called the source an 'LGM' (standing for little green men). She spent months eliminating the moon, passing satellites, and any terrestrial source. Finally she worked out that the stellar Morse code was more likely to be coming from an old star that had shrunk down to a few miles across and was whizzing around and sending out flashes like a lighthouse, as punctual as a ticking clock. It was a major intellectual breakthrough. (They gave the Nobel Prize to her supervisor, though.)

Radio astronomy was still in the game, it was part of the arsenal. And yet now even the giant dishes – like the original neoclassical observatory – had begun to seem a little archaic, almost quaint, verging on obsolete. Slightly passé. Another historical monument. From my point of view anyway.

All astronomy had run into one extremely big hitch, as Juliet had shrewdly implied. Whether optical or radio, astronomers relied on picking up electromagnetic waves of one sort or another, photons from far-off places. So long as there were photons around, they could see anything and everything, given a powerful enough telescope (or radio telescope). All you had to do was keep amping up the technology. But clearly, if you run out of photons, you are stuffed. You have nowhere else left to go. You have nothing left to detect. You run into the buffers. And that is just what had happened: astronomy had hit the limit of its powers, somewhere around 380,000 years after the creation of the universe. That's when the photons stop or, looking at it from the point of view of the evolution of the universe, when they start, when photons and electromagnetic waves emerge for the first time from the amniotic plasma. With the benefit of forward-facing chronology, the point at which, in effect, God said: Let there be photons. Or *photon*, singular: the first photon, a glimmer, like a single moonbeam, the first spark out of which everything catches fire. Thereafter, the faint red glow of ionized gas, and eventually (millions of years later), the first starlight.

But before that, the pre-photon era – nothing, electromagnetically speaking. The whole universe is hidden inside an impenetrable fog. A veil comes down and darkness descends and everything is without form and void and only blindness remains. Thus any telescope – even if you thrust it up into space and sent it into orbit, like the Hubble – was always necessarily going to fall nearly 400,000 years short of my goal. It was like that cruel moment in Kafka when the doorkeeper tells you that this door is open just for you – and then slams it shut in your face (and then you die). There was no way into the inner sanctum, the boudoir of creation. I had hit what the physicists call the Dark Ages.

I remember when I first fully realized this. I was in New York. I had been counting on pinning down the origin of time and space, maybe not now but soon, one of these fine days. There was no rush, I had all the time in the world, I could afford to be relaxed about it, the outcome (so I thought) was inevitable. And then: I discovered that there were no more photons available to be spotted. There was a cut-off point, a horizon. I'd always imagined, in my dreams, that there would be a few at least roaming around, early electromagnetic wanderers thrusting out into the great emptiness, some passing particle or waveform that a telescope could finally hang its hat on. But no, nada. I'd had dinner the night before with Avis Lang in Society, a café on the east side of Morningside Park. She lived around the corner from the Natural History Museum, she was an editor at a science magazine and collaborating on a project with Neil deGrasse Tyson, the black astrophysicist, and she was intent on shooting me down in flames.

'Andy, you are living in a dream-world,' she said. What I wanted was about as likely as 'the gold at the end of the rainbow'. She was really drumming it in with a kind of sadistic pleasure. The point was: even if there was any gold there, I would never ever see it, there was nothing to see it with, nothing but darkness and the *tohu-bohu*. I was sunk. She reckoned she had the director of the New York Planetarium, Stephen Hawking, and a few other big shots on her side. I ought to give up and go back to surfing, that was her line. She was a more grown-up, tougher New York Juliet.

I laughed it off and tried to think of other things for a while. But around 6 p.m. the following day it really hit me. The dream was dead. There was no getting away from it. I sat in a bar on Broadway, up near Columbia, and sheltered from the clear night sky and all the skyscrapers pointing relentlessly up at it. It seemed to me, at that point, that the heavens were meaningless after all. I couldn't bear to look any more, it was just an abyss, the void, empty. There were no answers there. I was never going to get the box open and take a look. No short cuts, no trapdoors, no visible and demonstrable truth at all. I had been wasting my time. I chose to look into the depths of a large glass of wine instead, as if that contained some kind of truth. But the truth, it seemed to me now, was surrounded by an invisible brick wall, it was locked up and somebody had thrown away the key. Cosmic censorship had been ruthlessly applied.

It was only then, running up against those buffers, that I realized that for a long while I had been thinking: I'm going all the way! I had been counting on it, without putting it in so many words. It was a kind of cosmic (over-) confidence. But I had just been fooling myself all along. I knew: there was no chance of going all the way, there never had been. In other words, extrapolation, theory, speculation – well-informed speculation, but speculation nevertheless – would take over again. Contrary to Galileo, the universe was *not* 'open to our gaze', not entirely anyway. After (or before) 300,000 or so it was closed and no one was ever going to open it up again. The deconstructionists were right after all, and there was no real access to the origin, the great *arché*. Everything was hopelessly hazy and indeterminate after all. End of story. Finito. I sat there feeling sorry for myself, of course, but in some obscure way I felt sorry for the whole of humanity, who would never know – not really *know* – whatever the hell was going on. It was back to good old educated guesswork.

There was an irony in all this. When, in 1992, the satellite known as COBE (Cosmic Background Explorer) discovered 'wrinkles' in space, deformations in the cosmic microwave background, critical fluctuations in temperature (to the tune of 1 part in 100,000) that were like premonitory sketches of the galaxies still to come, dating from around 380,000 years after the Big Bang, the newspapers ran the story with the headline: 'THE FACE OF GOD'. (*Newsweek* had the more literate 'THE HANDWRITING OF GOD'.) I thought (later, when I came to consider it carefully): ha! Hype! This isn't the face of God, this isn't even the rear end of God, this is just another image of the absence of God. Unless you could rewind all the way to the beginning, you would never find God. God was as invisible in the year 380,000 as in the year 13.7 billion.

To be fair to the newspapers who carried that headline: they were only quoting George Smoot, one of the two 'authors' – Smoot and Mather – of the project. Like Unc, Smoot had carried out a lot of balloon experiments which ended in disaster when the balloon blew up or was blown away by a hurricane. He switched to a U-2 spy plane, but finally – after years of disappointments – he managed to get his satellite launched right out of the atmosphere on the back of an ageing space rocket. When he saw the first pictures that the satellite produced of this alternative map of the heavens, with the subtle but crucial variations in the background radiation, Smoot exclaimed: 'It's like seeing the face of God!' On the other hand: he would say that, wouldn't he? The newspapers didn't have to buy it. Smoot's less headline-worthy version, delivered in a press conference, was: 'We have observed the oldest and largest structures ever seen in the early universe. These were the primordial seeds of modern-day structures such as galaxies, clusters of galaxies, and so on.' Smoot and Mather had come up with a powerful confirmation of Big Bang theory, but all the talk of God was still a massively over-inflated metaphor. And to be fair to Smoot and Mather, they were awarded a Nobel for their work in 2006, so a lot of people were fairly impressed with what they had come up with. I was impressed too, but it was just not enough. Like Bill Healey in Hawaii, I wanted *more*. From my admittedly extremist angle, it was everything or nothing. Half-measures, almost by definition, were just not up to the mark.

Most physicists are reasonably content with their success so far. They describe how much they have achieved with the following slightly suicidal metaphor (which I am stealing from Alan Guth, originator of 'inflation theory'): if you started off at the top of the Empire State Building and you were aiming at getting all the way to the bottom, you had already gone all the way down, except for the very last inch. You hadn't quite crashed into the sidewalk yet. It was like the old joke about the guy who jumps off the top of a very tall building, a hundred storeys high. As he goes flying past the twentieth floor on his way down, someone yells out to him, 'How's it going?' and he replies: 'So far so good.'

You could turn it around (up is down and down is up) and say, more heroically: we are reaching for the pinnacle and we are now within a centimetre or two of the summit. To most people that looked like a job well done. To me, in all honesty, it didn't. You could say that we were homing in on the truth, like a detective pursuing a 'promising line of enquiry'. The perpetrator was just out of sight around the next corner. But the reality was: I felt as if I was going to have to put up with a kind of halfway house, with some extra theory on top to make up for the lack. From my point of view, 380,000 years, however you sliced it up, was still an awful lot of potatoes.

WMAP (Wilkinson Microwave Anisotropy Probe), the successor to COBE, some 35 times more sensitive, has produced and continues to produce ever more detailed maps of the cosmic background radiation in the early universe. The maps can be easily consulted on their website. WMAP came up with the figure of 13.7 billion years for the age of the universe that I refer to in this book (with a possible error variation of 0.2 billion years). The Space Telescope Science Institute went so far as to announce that 'We might have seen the end of the beginning.' Which was no mean feat, and yet: the beginning had to have had a beginning. And it was the beginning of the beginning that remained tantalisingly elusive. The search had uncovered *more*, but it had not attained its goal. I should have been satisfied, but in truth I was anything but.

Then Unc told me about grav waves. Small but perfect. They refreshed the parts that other waves couldn't reach. The dream lived.

15

I was looking for LIGO. But all I could see was endless wide open spaces and tumbleweed tumbling about in the wind. I was out of water, the sun was blazing down, my cellphone didn't work, large black birds were circling overhead, and I was starting to get hysterical. 'You bastards!' I let fly out of the car window. 'Why didn't you give me better directions?' I wasn't talking to anyone in particular, I was encompassing just about everyone I had spoken to in the last couple of days, including me. It was just that I had an idea that a piece of hardware incorporating two steel tubes 6 foot high and 4 kilometres long in an L-shape configuration ought to be visible for miles around. So I stopped concentrating when it came to the last leg of the itinerary. I didn't realize I was going to end up in the back of beyond, somewhere on the outer rim of nowhere.

'I go LIGO.' It was a stupid pointless rhyme that had been rattling around in my head for the last couple of hundred miles. Now it seemed stupider than ever.

I had flown up from Los Angeles to Portland, Oregon. I fondly imagined I knew the West Coast. I was at home there, I knew surfers all along the coast, even in the most northerly reaches of the west. In fact, I had a kind of home from home. Auntie Jenny (a rather remote relative, originally from Australia, but still) had split up with her husband of thirty-odd years and left Boston and gone to live in Portland. She picked me up at the airport and drove me back to her place. So recent was the move that she took a wrong turn and we started hitting mountains before she realized her mistake, turned around, went through town and out again to Lake Oswego. I'd hardly put my bags down before three of her sons converged on the house (arriving from Sydney, Seattle, and Portland itself), together with a couple of grandchildren, a boy and a girl, both of fairy-tale beauty, with Asiatic features.

'We're going to the zoo,' said Andrew (the kids' father). 'You want to come?'

'Well, I just blew in and I've got some stuff to work on for tomorrow.'

So we went and bought some sandwiches and went to the zoo.

Oregon Zoo was an overwhelming microcosm of life in all its diversity and abundance and stripy, variegated, bewhiskered, feathery strangeness: hippos, pandas, monkeys, hyenas, miraculously poised mountain goats, snakes, scorpions, a solitary polar bear, panthers, cheetahs, chimpanzees, lions, sea lions, cougars next-door neighbours with giraffes, bald eagles, assorted bears (snoozing), beavers and buzzards; a compendium of landscapes, savannah, arctic, Africa, Australia, Alaska, tropical rainforest, temperate zones; endangered species (a lot, possibly all of them); some rare and beautiful creatures I couldn't even identify; and, most numerous of all here, as in most places, human beings, who had brought all of the other species together and stuck labels on the cages.

We were looking in at them but they were also looking out at us. It struck me that we must look pretty strange and perhaps quite amusing to the inmates of Oregon Zoo. Maybe, from their point of view, we *are* the show (although only the monkeys seemed particularly entertained by it). Presumably, if an alien race were to come down, kidnap a few dozen earthlings, and stick us all in cages, they could find a few labels to put on us too: differentiated according to size, age, colour (eyes, skin, teeth), weight, dietary preferences, sexual orientation, brain power, and whether or not you have hair on your back. Come to think of it, we don't need aliens to do all that, we do it already. We are incurable taxonomists, no matter that so many of our categories are arbitrary to the point of brutal.

'Are you our relative?' Jenny's grand-daughter asked, turning from an amiable, well-fed family of crocodiles towards me.

'Yes, I am,' I said.

'What relation *are* you?' said the boy.

'Uncle,' offered his father.

'More like third uncle-in-law,' I said.

'Uncle,' they said.

I tried to think out precisely what my relationship was to the couple of kids, because they asked me. Even 'third uncle-in-law' was an absurd oversimplification, there was a grandmother in there somewhere, and a couple of in-laws, and in the margins of all of that a grandfather who had gone off and married a Chinese girl, and I was somewhere in the middle of it, so without a pie chart or a graph or sophisticated graphics there was no way of making sense of it all or, indeed, any of it.

And that was thinking only of the handful of human beings I actually knew who happened to be in Oregon Zoo on that particular Sunday afternoon in 2007. After that it got really complicated. But I knew that if you could trace it back a few generations, most of us would turn out to be connected, to have common ancestors, and it would start to look a lot less complicated. Keep going far enough back, and every single person on the planet would turn out to be 'family'. Keep on going, and all those different cages and labels, plumed squirrels over here, and the rare and mysterious Tibetan puma over there, the immense cascade, the chaos, of sheer multiplicity, from armadillos to zebras, would diminish and fade away and cease to have any meaning whatever. We would be reversing in the direction of simplicity, of the impersonal and the inorganic, far far away from the realm of suckling pigs, stray dogs, and mermaids.

The next day I could have gone jogging around Lake Oswego with Daniel or had coffee and a croissant with Nick at Starbucks. But I had a date with our lost common origins, so I turned down all offers and hired out a car from Enterprise and drove east. And kept on driving. I followed the curve of the Columbia River as it bent north. I left behind the city, I left behind Mount Hood, 11,239 feet and still snow-capped, I left behind plunging green valleys and densely wooded slopes and silvery waterfalls. Somewhere east of Arlington all of that clamorous, colourful existence started to coalesce into a monochrome miasma of dust. Which is when I became utterly and irredeemably lost. I was somewhere in the state of Washington. I knew I wanted to be on route 10, and there were still roads, but there were no more signposts. Only a landscape without form and void. So it was that I began winding down the windows and screaming. I don't function very well in the absence of signposts. I still needed some basic taxonomy to get me to my destination.

I pulled in at a passing gas station. I asked a cashier and a mechanic if anyone had any idea where LIGO was. Nobody had. I assumed everybody would know. Nobody had a clue what LIGO was, let alone where it was. Maybe they just didn't care. I think I must have been getting irate. I asked one woman with piled-up flame-red hair, who was just getting out of her car, on my way out. She gave me some transparent lie, 'Sure, honey, it's over there,' with an indeterminate wave at the great nothingness.

'Forget it,' I snapped. I'd already been 'over there' and I wasn't going back again.

'I don't want to be knifed,' she said.

'Do I look as if I have a knife?' I said.

'I don't know,' she said. 'Do you?'

I took a look at myself (as I used to be a short while before) in the rear-view mirror. Was I mad or was she? The mirror didn't say. I was wishing I'd gone for a jog around the lake after all.

About an hour later I ended up at the Center for Astronomical Research of Northwestern University, on the edge of Richland. It was completely the wrong place but at least they didn't look at me as if I was the man from Mars. They took pity, and gave me a glass of water, and steadied my nerves, and phoned up a woman named Marina. Marina, when she finally appeared, sent me away with a detailed map of the Hanford site. I followed an empty road past a well-guarded nuclear reactor. It occurred to me, as I went by, that this was probably the same nuclear reactor that people back in Portland had warned me to stay well away from on account of a history of massive radioactive leaks. One way and another I was guessing that LIGO didn't get a lot of visitors. Maybe they didn't really want visitors.

A few more turns then the road straightened out and kept straight all the way to infinity. I was beyond despair. I was content just to keep driving until I ran out of fuel and then I would get out of the car and stagger around for a few miles and then die. Then I saw it. Or them.

The architecture was almost unsurprising. At this distance, at the far end of a long, emphatically non-tree-lined avenue, it could have been a bunch of boxes, stuck at odd angles, blues and greens, a triangle, some glass, a few curves thrown in, nothing in the least out of the ordinary. But then, like a landing strip for aliens, or a more linear alternative to crop circles, two tubes, elongated domes, making a bifurcating path, like an L or two arms of a cross, burrowing across the desert, vanishing over my horizon. A divining rod on the scale of the cosmos. It must have been visible up in space and it could have been a signpost to wandering aliens, advertising our intelligence, but in fact it was more of a sketch pad for great extraterrestrial forces to announce their presence and leave a signature and an address. A cosmic visitors' book. 'Laser Interferometer Gravitational-Wave Observatory', the sign said. I had re-entered the realm of signs.

I parked the car and slammed the door, as if this had all been the car's fault (vowing to rent one with satnav next time), and walked through the front door. I was approximately two hours late. There was no one around. The place was deserted, unmanned. It looked like the *Mary Celeste* in there. Then I saw the huge cylinder, sitting in the lobby, like a very substantial ghost from our scientific past. Weber's bar.

It was only aluminium, but I felt as if I had just struck gold. The door was open, there were no guards, but it would have needed a very determined kleptomaniac armed with a crane to get away with this baby: I guessed it must have weighed a ton. A solid cylinder, massive, vast, implacable. The opposite of delicate. You couldn't knock it off its plinth with a sledgehammer. But it was a Ming vase among scientific artefacts. Perhaps locked inside it were the first secrets of our non-electromagnetic history. The secrets of the beginning of any history.

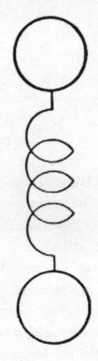

Figure 8.1

Weber's diagram for the Weber bar

16

Joseph (originally 'Jonas') Weber was pleased, even ecstatic, but probably not all that surprised when he first detected gravitational waves ('grav' or sometimes 'gravity' waves for short) in the middle of 1968. After all, he had had a lot of experience of waves, one way and another. He served in the US Navy for a number of years. In fact he was an undergraduate at the US Naval Academy (awarded a Bachelor of Science degree in engineering in 1940). Theory alone was never going to be good enough for him: he had to *do* something with knowledge, make it work, apply it in a practical way. He became a radar officer. During World War Two, when he was only in his early twenties, he rose to the rank of lieutenant commander. His ship, the aircraft carrier USS *Lexington* (also known as 'Lady Lex'), only narrowly escaped catastrophe, steaming out of Pearl Harbor on 5 December 1941, two days before the fatal Japanese attack.

Weber had a second close shave. The *Lexington* was heavily involved in the Battle of the Coral Sea in 1942 and sank the Japanese carrier *Shoho*, but was seriously damaged in the encounter. Eventually, the ship started going down. Weber managed to get off along with most of the crew and he watched with a kind of detached awe as the ship slipped down beneath the waves, glowing incandescent as it did so. Like anyone who comes this close to death – not once but twice – Weber was convinced that he had been saved for a reason, that he had to achieve something significant with his life. The experience of a ship riding the ocean waves provided Weber with his model of what a terrestrial detector ought to be like when it came to grav waves. The 'Weber bar', as it became known, ought to bob up and down (ever so slightly) as the (almost imperceptible) waves passed beneath it. All you had to do was measure the displacement.

But it took a while for him to home in on what would become the focus of his life. Born in New Jersey to a Lithuanian father and Latvian mother, the youngest of four children, he was already a radio ham operator by the age of eleven. He bought his own radio and books on electronics by working as a golf caddie during the Depression for a dollar a day. After the sinking of the Lady Lex, Weber went on to command a submarine chaser, protecting Atlantic convoys, and landed in Sicily in 1943. After the war, he worked as head of electronic counter-measures at the Bureau of Ships for a while, but he was hired in 1948 by the electrical engineering faculty of the University of Maryland. Only 29 and already a full professor. Such was the chaos of the post-war years that he became a professor first and did the doctoral thesis afterwards, getting his PhD (for work on the microwave inversion spectrum of ammonia) from the nearby Catholic University of America in 1951.

It was around the same time that Weber had one of his greatest insights: he realized that the A and B coefficients in Einstein's theory of stimulated emission could be exploited to generate amplified electromagnetic radiation, thus leading to (as we now know them) masers and lasers ('light amplification by stimulated emission of radiation'). He envisioned beams of pure microwave (hence '*m*aser') energy leaping across space. Perhaps, in the manner of the science fiction magazines of the period, he foresaw maser cannons blasting enemy planes out of the sky, punching a hole clear through the metal chests of marauding robots. In 1952 he was the first to speak about the maser principle, in Ottawa, at the Electron Tube Research annual conference. And he published what was the first open-literature paper on the theory of quantum electronics. But he had no research funding.

Others, like Charles Townes in New York, and Nikolai Basov and Aleksandr Prokhorov in Moscow, stole ahead and developed the first functioning devices, the real thing. *They* received the Nobel prize. Weber, on the other hand, received an award from the Institute of Radio Engineers for his 'early recognition of the concepts that led to the maser and the laser'. It must have been frustrating, galling even, to see his ideas slip away – taken away – from him and turned into reality elsewhere. 'I was only a student in a way,' he said. 'I didn't know how the world worked.' Well, now he knew. He wouldn't let it happen again. The next time he would make sure that he *owned* the apparatus too. But, after this bitter experience, he was sick of electromagnetic phenomena. To hell with anything to do with light, he thought. He'd had it with light. If he wanted to really see, he needed to go beyond light, in the direction of gravitation, for example.

In the mid-fifties, Weber had a year off, a research year, known in the business as a sabbatical leave. Perhaps he was still getting over the disappointment of getting outmasered. He took himself off to Holland, where he studied at the University of Leiden, and then returned to the States and continued at the Princeton Institute for Advanced Studies, where Einstein had been a resident scholar. It was during this time that he took up the cause of general relativity, influenced by John Wheeler and Freeman Dyson. Dyson had been calculating the gravitational waves that ought to be emitted by the collapsing stellar core of a supernova. He worked out that the gravity wave signal would be a lot stronger than previous calculations had suggested.

By 1961, back at Maryland, fired up by his new passion, Weber had switched from engineering to physics and published a paper in *Physical Review* and an elegant monograph, *General Relativity and Gravitational Waves*, full of talk of negative mass and antiparticles and long-winded concatenations of Greek symbols. But he remained an engineer within physics and was impatient with anything that remained at the level of pure theory, like a mathematical equation. What was the point of an equation, Weber instinctively felt, unless it had an application and a real solution? You had to be able to take out the abstractions and the imponderables and slot in the particular and the known. 'The equations of general relativity,' Weber wrote, 'have solutions which appear capable of giving a description of the universe and its evolution.' Forget the spaceman paradox: there are no 'real' paradoxes. No matter what 'it' was in the sentence 'it is an empirical fact that *it* has never been observed' – *it* was unworthy of further serious consideration, unless, that is, someone was going to go ahead and observe it. He admired not just Einstein, but equally the 'ingenious experiment' of Eötvos (testing the ratio of inertial mass to weight). Perhaps this was why he was led into the wave field, 'the most fundamental field in physics': he wanted to overcome mere theory.

Like Joe Weber, I saw gravitational waves as the answer to all my problems. They were being generated at the very beginning of the beginning. For every action there is a reaction. For every interaction there are waves. Nothing – especially nothing as big as the Big Bang itself – could take place without generating waves of gravity, albeit rather small ones, frissons, ripples, 'perturbations', fine corduroy lines in the fabric of space-time, so that the great vastness is vibrating, pulsing, shuddering, breathing in and out. It was as if the universe was not an empty, cold, silent, inhospitable kind of place after all: it had a heart, or hearts, at any rate, beating out a rhythm, pumping out invisible tides across the foaming firmament. Travelling at light speed, waves would go storming right through the original plasma, the formless, chaotic cocktail of primeval gases and particles, the formation of the first galaxies, the emergence of planets, the era of fire, of water, of air, of earth. They were pure alchemy. There was no 380,000-year cut-off point. They would keep on rolling all the way through the evolution of plant life and animal life on earth. Until finally *Homo sapiens* – in the shape of Joe Weber and others – was in a position to detect them. All you had to do was catch one and ride it into shore.

So it was that Weber became the first man to take up the challenge implicitly thrown down by Albert Einstein. At first Einstein refused to believe his own (general relativity) equations. In 1936, Einstein and Rosen – the same collaborator with whom he tried to abolish entangled particles – submitted a paper to *Physical Review* asking the question 'Do Gravitational Waves Exist?' Their decisive answer was: No. But the editor sent it back. A referee had spotted some crucial errors. Einstein was insulted and withdrew the paper. In the end – the idea was almost incredible – he was forced to retract. He came round to accepting the implications of his own equations. But he had a fall-back position. Gravitational waves were real, he admitted, and they were 'cylindrical', but it made no practical difference, no one will ever be able to measure them, they are just too small by the time they reach Earth. You could see light bending around very massive objects, you could measure it. You could even detect relativistic time effects in our own solar system. But gravity waves, bearing messages from the origin of the universe, documenting catastrophic supernovae and the collision of mighty stars – they were just too subtle, too imperceptible, for ever beyond measurement. Think of a big wave, Hawaiian-sized, Waimea Bay for example. Now scale that down to, say, a gentler English wave, gently unfurling on the beach. Now divide by two, and keep on dividing by two, to the point that you have nothing left, nothing you can see. Imagine a wave that increased the height of the ocean by a mere 10^{-18} metres for every metre of water. 1 over 1 followed by 18 noughts, Which looks like this: 0.000000000000000001. Or 1/1,000,000,000,000,000,000th.

As a fraction of a length. Not much, is it? It doesn't appear on any tape measure. Less than a thousandth the width of a proton if you happen to be measuring the Empire State. On a scale from here to the Sun it is the size of a germ. Calculating the distance from here to Alpha Centauri, you would be looking for something the thickness of a human hair. Forget it.

This was the crucial back-of-the-envelope equation (which Greg Mendell actually scribbled on the back of an envelope for me):

$$h \sim \frac{4\pi^2 G}{c^4} \times \frac{Q}{r} f^2$$

where h = amplitude (ie the 'height' of the wave) and Q = quadrupole moment and G = Newton's gravitational constant. At frequency f. (In fact I see he didn't write ' = ' but ' ~ ', meaning it was all rather approximate.) But the really significant factor is the bit at the bottom, the divisor: c – the speed of light – to the power of 4: $c \times c \times c \times c$. Remember that the speed of light is already an exceptionally large number. Any equation with that many cs on the bottom was going to produce an infinitesimally small answer no matter how you sliced it up. In a way, you could compare these waves to Unc's entangled particles: what they had to say might be deeply instructive, but unfortunately there was no practical way to take a look-see. They were untouchable.

But to Joe Weber it was obvious: if these things were real, then it had to be possible to find them. After all, he had done it with radar. It was logical. If you could build an antenna to pick up electromagnetic waves (and the entire post-war field of radio astronomy told you that you could), then it ought to be possible to build one to pick up grav waves likewise. It all boiled down to coming up with a good enough detector, a sensitive enough aerial. Every prediction could be tested. The physics was already in place: all that remained was the engineering. It was doable.

Weber filled four whole 300-page notebooks with drawings and equations. He spent years dreaming up brilliant inventions to catch those elusive waves – and, one by one, he had to junk them all. For a while he wondered if the entire planet could be recruited as a detector. The Earth could be plucked like the string of a gigantic harp, sounded like a mournful bell, thus generating measurable oscillations. But background noise knocked the idea flat. He persuaded NASA to take a gravimeter into space on Apollo 17 to measure the vibrations of the Moon. Joe Weber was the great pioneer, the pathfinder on the long road to the first wave. 'My philosophy,' Weber said, 'was to act like Galileo: build something, make it work, and see if you find anything.' Maybe the universe was not exactly 'open to the gaze' any more, as Galileo argued, but it was still open to inspection: you just had to come up with a more subtle tool than a telescope.

He also had the more recent example of Heinrich Hertz to inspire him. In 1864 James Clerk Maxwell came up with the theory of electromagnetic waves to explain the affinities and liaisons between electrical and magnetic fields. It was a good theory, but it took Hertz, in 1880, to devise the apparatus (which would become known as an 'oscillator') that would show that such waves really existed. Hertz had only two rods and a gap: when a spark jumped the gap, it proved the presence of radio waves. And he was able to show what Maxwell had only hypothesized, that radio waves travelled at the same velocity as light, so that electromagnetism was in fact a form of light (or rather that light was just one part of the electromagnetic spectrum). And all of these took the form of waves. In the same way, all Joe had to do was pull real gravitational rabbits out of Einstein's theoretical hat.

Weber never had any illusions: he knew it was going to be tough. You had to be able to measure a displacement smaller than a nucleus. Fellow physicists thought it would take more than a century of experimental work to get close. In public Weber issued self-deprecating statements. 'The probability of success under these circumstances ha[s] to be regarded as very small.' No one had ever done it before. But that was what made it so exciting: to get there first, ahead of the field, to land the smallest-ever fish. It was the exact opposite to the classic angler's tale: estimates of size had to be radically scaled down, contracted, shrunk. Small was beautiful. He was climbing a mountain of minuteness. The late fifties and early sixties was the golden age of the transistor, which replaced the clunky old vacuum tubes, and the portable transistor radio. It must have seemed only natural to think in terms of the miniaturization of signals as well. Gravitational waves were the transistor radios of astronomical observation, beeping out their cosmic messages. So far they had managed to slip through the sieve. Weber set about devising a finer sieve, the finest ever invented.

Call it a mousetrap, if you will, for trapping gravity waves, far far smaller than any mouse. What Weber finally came up with – surprisingly, at first glance – was a large solid cylinder made of aluminium. Around 2 metres long and 1 metre wide, total weight around 2,600 lbs, it looked like a steam engine without wheels.

Around its waist was a belt of piezoelectric crystals, designed to measure the oscillations after the bar had become 'excited' by the passing waves. The crystals had the property (first noticed by Pierre Curie in 1880) of generating a voltage when squeezed. The output could be read on a chart similar to that used to record earthquakes. The waves would be like tiny seismic shocks hitting the bar. As the theory of neutron stars and black holes began to crystallize in the sixties, Weber could legitimately envisage the whole universe shuddering, vibrating, under the impact of huge and violent phenomena, stars colliding or collapsing, black holes merging, vast and violent forces knocking ten bells out of each other. His 'resonant mass' detector was designed to pick up the shock waves (and it would be the right size for frequencies of a few thousand hertz, the kind that would emanate from such mighty catastrophes). The pen jumping up and down on the chart would show their rise and fall, the throbbing signature of some distant event, the heartbeat of the universe. Ultimately, potentially, the most distant, the most remote in time of them all.

When, in December 1968, Weber's pen started inking in waves – like a seismometer of outer space – and kept on scratching out signals in the months to come, Weber dashed off an article to *Physical Review Letters* and he got up and gave a speech announcing his discovery to a big audience at a relativity conference in Cincinnati. Cue applause and hurrahs. The standing ovation. In another age top hats would have been flying.

For years, all the way through the sixties, conversation among general relativists had invariably come around to one recurrent question: 'Has Joe spotted anything yet?' Perhaps there was an edge to the question, an abrasive rasp, a mocking undertone. It was almost a standing joke. Now Weber was able to reply: 'As a matter of fact – yes.' The sceptics had been routed. Weber's classic paper, 'Evidence for Discovery of Gravitational Radiation' (*Physical Review Letters*, Volume 22, Number 24, 16 June 1969), demonstrated that he had duly picked up the expected vibrations. Mission accomplished. QED. Weber had cracked it. Weber had two bars in separate locations, like Galileo's two men on separate hilltops, but now they were 1,000 km apart, one at Maryland, the other – the control – at the Argonne National Laboratory, Illinois, south-west of Chicago. Their charts showed the same range of deviations, their pens were jumping at the same time, ergo this couldn't be accounted for by some localized energy source. It was the kind of coincidence that would occur naturally only once 'every 7×10^7 years'. Weber had ruled out cosmic-ray showers and seismic disturbances. Everything had been shielded to block out extraneous interference (as early as 1961 he had said 'the apparatus must not respond to earth vibrations' and spoken of 'acoustic filters'). He wasn't too sure about exactly where they came from or what they signified – a supernova? a pulsar? – but Weber was confident he had registered the first physical evidence of the existence of gravitational waves. He had finally picked up the pulse of the universe. His resonant mass had resonated.

1968: a period of revolution and enlightenment. Celebrated with Molotov cocktails, the boulevards of Paris were witnessing the 'Evénements'; but Argonne could boast of cosmic 'events' that would eclipse all others in their significance. The doors of perception had suddenly flown open. Or at least, as Weber put it, perhaps consciously echoing Huxley, 'a new set of windows for studying the universe'. The key to all mythologies. It was a quintessential sixties phenomenon, Weber's seemingly inert bar was tuning in and turning on, getting with the vibe, throbbing to the beat of the universe.

Weber had finally proved it could be done, in the teeth of all the scepticism, even from Einstein himself. Hearty congratulations all round. Headlines. Photographs of the man with the piercing eyes that seemed to see the unseeable. That Nobel was surely in the bag.

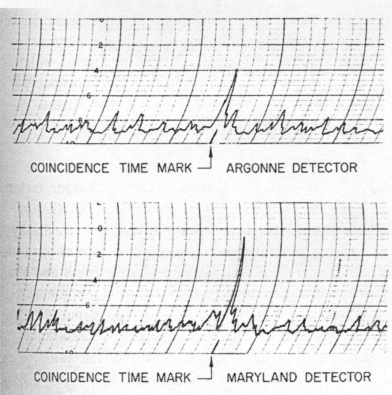

COINCIDENCE TIME MARK ⌐ ARGONNE DETECTOR

COINCIDENCE TIME MARK ⌐ MARYLAND DETECTOR

FIG. 2. Argonne National Laboratory and University of Maryland detector coincidence.

Weber's Waves

17

In Cambridge, England, Stephen Hawking was excited by Weber's discoveries. With his co-author Gary Gibbons, he set about analysing the kind of signal that Weber's detector could be detecting. At this time Hawking hadn't written his *Brief History of Time*; he had barely completed his PhD and was a humble research fellow at Caius College. But he had already been diagnosed with some form of motor neurone disease, and didn't think he had long to live, so he was in a hurry to get all the major problems sorted out before he died. Gravity waves could be the answer to everything, a short cut to a total vision of cosmology and the 'mind of God', not just the face.

Hawking hypothesized that the probable source of Weber's waves was newborn neutron stars, elementary pulsars. There was a problem though: the Maryland bar was giving out signals of significant events loud and clear every day, propelling the pen this way and that on the page, like a powerful spirit from the beyond shoving the glass around on a Ouija board. For the kinds of energies that were being detected, the stars had to be (or have been) located within roughly 300 light years of Earth. But there couldn't be that many neutron stars that close. So maybe they were coming from closer to the galactic core, more like 30,000 light years distant, as Weber himself suspected. But if so, then the energies being delivered would have to be a hundred times higher to make up for the extra distance involved. According to Hawking's sums, every event, as seen by Weber, would have to correspond to at least one sun's worth of mass and energy being converted into waves. But Weber reckoned he was picking up signals on this scale daily. Another star going up in smoke, *every single day*? A sun a day for billions of years? A lot of days, a lot of suns. Hawking calculated that, if this were true, then there would be practically nothing left of the entire galaxy: our neighbourhood, the cosmic *quartier* that was the Milky Way, should already have gone up in smoke. Therefore there would be nothing left to produce the signals. There was not enough stuff in the local universe to produce that number and size of waves. And yet, there – apparently – they were: the waves, the signals, the pulses. A message that implied the annihilation of everything and the impossibility of messages. A paradox. If it was true then it was false. Not that Hawking, who was becoming preoccupied with black holes and singularities, was averse to paradoxes. But it was worrying all the same.

Waves too big? Well, replied Weber, that was just a big swell, bigger than the average. Call it beginner's luck.

In the summer of 1974 Weber went to the Fifth Cambridge Conference on Relativity – the other Cambridge – held at MIT. It was like walking into the OK Corral of astrophysics. Waiting for him there, ready for the showdown, was Richard Garwin. Garwin was roughly the same size as Weber, slightly lighter, with thinning hair, but he was a physics heavyweight. He had been involved in designing the first hydrogen bomb (a thousand times more powerful than a mere atom bomb); he worked for IBM; he was on the President's Science Advisory Committee. He wasn't an astronomer but then neither was Weber. And he was something of a rebel who took nothing at face value.

Garwin was frankly dubious about Weber's results. At the end of the sixties he decided to settle the issue once and for all by building his own detector. He figured that if he could build a nuclear bomb he could damn well build a gravitational wave detector – how hard could it be? He put together a 260 lb miniature detector, based on Weber's design, in six months flat. In 1973 it picked up a pulse. So – Garwin had to admit it – maybe Weber was right after all. Respect to Weber.

Until Garwin's friend David Douglass came along and told him it was all down to a programming error. Weber was seeing a signal in what was pure noise, picking up phantom pulses. He even claimed to have detected another 'coincidence' between his devices and another one run by Douglass in Rochester. But Douglass pointed out that they operated in different time zones: Rochester was Eastern Standard Time, but Weber's labs worked with Greenwich Mean Time. Time difference: 4 hours. So how could the apparent coincidences in the data really coincide? Weber must be slanting the data, tilting the pinball machine to suit his own purposes. He was, in a word, cheating. Not quite a hoax, but it was close. An hallucination? What was this guy on? It was more like Carlos Castaneda than hard science. Garwin was annoyed even to have put in the time to check. Six months of hard labour and all along Weber was as good as making it up, improvising like some jazz musician, beating out his own tune on the bar, tapping his foot in time to a rhythm that was in his head. The point was: nobody else could hear it.

Garwin let it all out at the conference, tearing into Weber. He made sure that everyone else could hear what he was saying. He was calling the supposed signals 'Pathological Science'. The great happening had never really happened. Weber was robustly built.

He had retained the mindset of the fitness fanatic from his years in the Navy. He jogged and swam every day and went rock-climbing along the Potomac in Carderock. He had a sort of motto about this, emphasizing the need for physicists to stay fit and healthy: 'You can't do physics when you're dead.' His iron-grey hair seemed to stand to attention. He was damned if he was going to let any pipsqueak bomb-crazy geek from IBM try to stiff him. Bertrand Russell once said that when it came to doing philosophy, the key question was: 'Can I take him or can he take me?' Weber and Garwin felt the same way about doing physics. It was machismo with math. Neither man was going to back down. They started slinging figures at one another, then graphs and computer printouts, on the stage at the front of the lecture hall. That didn't resolve anything. So the two of them got out of their seats and came towards one another with their fists clenched, their pockets bulging with pens and calculators. They were going to settle it finally, one way or another, man to man.

A cane came between them, waving about nervously. The cane belonged to Philip Morison, who had been stricken with polio as a boy. He was the moderator of the meeting and he was having to work harder to moderate the two men than he could ever have expected. The sight of the disabled Morison urging moderation upon them may have had some effect; that and the other physicists who came up behind them and held them back till they cooled down. Weber and Garwin resorted to an exchange of letters in *Physics Today*: Weber stood his ground, the signals were real, the waves were real, everything else was footnotes; Garwin came up with a computer simulation in which he managed to magic up a 'signal' out of pure random noise.

But the wave had surged to a flood. Despite the Hawking–Garwin axis, Weber bars proliferated: there were experiments going on all across America, in Russia, Australia, Chile, Scotland. The search continued unabated. They went down abandoned mines; the Soviets thought of shooting a dumbbell-shaped rotating detector into space. They tried replacing the bar with hollow squares, hoops, even horseshoes (for luck). They made them bigger and better, Bell Labs had one weighing in at 4 tons. Nothing, nowhere, except for a few earthquakes and secret nuclear tests. Nobody could detect those waves. Only Weber. 'You're just doing it wrong,' said Weber. They tried to do it right. They tried to up the sensitivity by taking the temperature right down. Still there was nothing definite and unambiguous such as Weber had picked up. In a word, nobody else could *replicate* the experiment.

Non-replication was the kiss of death in scientific circles. In a way it was the opposite of art. When Leonardo da Vinci came up with a new painting – or Gauguin or Picasso – people would say, 'That is a Leonardo!' (or a Gauguin, etc.). The point was that nobody else could do it in quite the same way, and if you did, then you were a fraud and the painting was a fake. When Galileo – or Joseph Weber – came up with a new observation then, in the contrary way, everybody else had to be able to say, 'I can do that too.' If, in fact, nobody else could reproduce what Weber had done, then it was just 'a Weber', merely a painting, a canvas of what was in Weber's mind, but not objectively out there and accessible to all. Tony Tyson of Bell Labs, for one, was furious with all the negatives. It seemed like everyone was spoiling for a fight. He denounced his own detector as 'the most expensive thermometer in the world'. Based on Weber's data, he complained, 'there should have been enough energy in other forms, such as electromagnetic waves, to knock your socks off. All you should have needed was a pair of binoculars.'

Weber resolved to stick to his guns. He wasn't about to cave in. It was like the war all over again. No surrender. To hell with Garwin, Tyson, Hawking, and everyone else, from Scotland all the way down to Chile. Not to mention a guy called Kafka in Munich.

He was an insect. Hell, they were all insects! Cross them off the Christmas-card list. So what if that made him a maverick? He knew more than the lot of them put together. Weber had been working on waves for more than ten years now, designing, building, calibrating, testing, tuning. He kept careful cross-referenced logbooks of everything. These signals had to correspond to something, they couldn't just pop up out of a clear blue sky. The pen had spoken. Let the whole universe fluctuate, but Joe Weber would not. He was as solid as an extremely solid aluminium cylinder. He knew he had detected something, but what exactly? Pulsars? He looked into the possibility that he had picked up solar neutrinos and other exotic entities. And he was willing, too, to toughen up the hardware. He added on diaphragms (more sensitive). He took the bar and plunged it into a bath of liquid nitrogen. That ought to calm down any inadvertently overexcited particles. Then, just to be on the safe side, he locked the whole thing away inside a vacuum as well. The bar was, for all practical purposes, dead to the world, but, conversely, all the more alive to signals from afar.

There was a residual problem though: the atoms that the bar itself was made of. They were liable to jump around, randomly, at the behest of quantum unpredictability. So they were capable of producing signals too. But what the hell were you going to make it out of, if not atoms?

18

obert L. Forward was Weber's PhD student. The L. stood for
'Lull'. He had parents who thought of everything in terms of
waves, of peaks and troughs, sets and lulls. Like Weber he had a
big quiff of dark hair and glasses with thick frames that put you in
mind of Buddy Holly. He had done his undergraduate studies in
physics at Maryland and then gone west, to UCLA, to take his
master's degree in 1958. But the prospect of working under Weber
drew him back to Maryland to study gravitational physics. He
completed his thesis, 'Detectors for Gravitational Fields', in 1965.
Forward respected everything that Weber had done. He thought
Weber had not just vision but guts too. He had been involved in
the construction of the bar under Weber's supervision. His
philosophy, from the very beginning, was (as he put it in his
unfinished autobiography) 'to work on problems that other people
consider impossible'. There was a natural match between him and
Weber. But when the bar started to hit a brick wall, Forward tried
to think his way around or over it, rather than trying to bulldoze his
way through it.

Forward, more than anyone at Maryland, demonstrated that what the quest to discover gravity waves needed was imagination. Scientists are not down-to-earth realists. Far from it. They are rather the opposite, yearning to float free of constraints, to ride out into the unknown, cowboys herding invisible cows. Lee Smolin said (in *The Trouble with Physics*) that physicists need to be part of 'an imaginative community'. Forward went one better and ended up becoming a writer of science fiction, drawing on his hard-won knowledge of the universe. He was working as a post-doc at the Hughes Research Laboratories in Malibu, California. In his spare time, in the early 70s, he wrote a story for *Analog* about a spaceship that came too close to a collapsing black hole and was damaged and nearly destroyed by the gravitational waves the hole emitted in the course of chewing up surrounding space. Jerry Pournelle, who had also been thinking about the lethal consequences of a huge gravitational tsunami, like Krakatoa going off, read it and was impressed enough to put in a personal call to the author.

Pournelle was about the same age as Forward (in fact a year younger, born in '33 to Forward's '32) and they had both served in the Air Force during the Korean War (Forward had specialized in radar defence and guidance systems). But he was already a seasoned science-fiction writer and essayist. Forward invited him to the lab and Pournelle took his friend Larry Niven along (they would eventually collaborate together on stories like *Inferno*, *Lucifer's Hammer* and *Oath of Fealty*). Forward gave the pair of them the grand tour of the Hughes lab and a blackboard talk on all the stuff he knew that might fit into science fiction. Like particularly intent students, Pournelle and Niven sat and listened with growing wonderment and admiration as Forward developed his idea of laser-powered lightsails for the purpose of interstellar travel (which they would ultimately weave into their *The Mote in God's Eye*) and his theory of miniature black holes, only millimetres in diameter, which would appear in the story 'The Hole Man'.

After that things snowballed. Pournelle was in charge of the Nebula Awards that year – the premier prize for science-fiction writers – and invited Forward to give the same talk at the awards ceremony. The editor of *Analog*, Ben Bova, heard about the talk and asked him to turn it into a 'science fact' article. Pournelle and Niven were by then running a semester at UCLA on 'Science and Science Fiction'. Each Thursday evening there would be a one-hour lecture by a scientist, followed by a one-hour lecture by a writer, and finally an hour of class discussion. One Thursday Forward was the scientist and Niven was the writer. Forward talked about what it would be like to live on the surface of a (very rapidly spinning and extremely dense) neutron star. He thought that it would be challenging to write a story about aliens and humans where the difference in the sense of time between the two was roughly on a ratio of a million to one. At the cocktail party at Niven's nearby home later that evening, Forward agreed to write up everything he knew about neutron-star physics as the basis for a novel. A month later, he duly delivered the text to Niven. Niven was busy. 'I tell you what,' he said. 'Why don't you write the first draft of the story, then I'll come in and we can collaborate?' Forward thought that sounded OK. 'Look,' said Niven, 'I know you are a scientist, but don't worry about it. You will get the facts right and the story will be dull and pedantic, but that's all right. I can fix that. I am a writer.' They agreed to split the book 50:50. Forward went away and produced the first draft of what would become *Dragon's Egg* (1980), his first novel.

He brought it back to Niven. Niven was still busy, collaborating with Pournelle. 'Why don't you go ahead and finish it?' he said. Forward finished it. To some extent Niven had been right: every publisher who saw it said, 'This needs rewriting by Niven or Pournelle.' Lester del Rey, another writer, read it and liked it but sent him an 11-page single-spaced critique. 'Make all these changes,' he said, 'and I'll read it again.' Finally his wife, Judy Lynn, a publisher (Del Rey Impact), bought it. Forward had jumped the gap between 'scientist' and 'writer'. It was like he had started off as Unc, my twin brother, and ended up as me (although a more scientifically literate me).

Dragon's Egg is known as a classic of the genre. It has characters (the Cheela) who are the size of a sesame seed but the same mass as humans, who evolve from primitive tribal wars to futuristic technological sophistication in the space of a few days, and refer to earthlings, affectionately, as 'The Slow Ones'. The human astronauts orbiting the star fear that 'by the time we can think of anything to say, the [alien] who asked the question down there would have died'. From the point of view of the Cheela, trying to have a conversation with a human involves long stretches between syllables when they have to go off and take an extended vacation. Every time the astronauts go to sleep the Cheela shoot through another millennium of speeded-up progress. From being far behind humans in intellectual development, they become so far ahead in a day or two that the Cheela have to stop communicating with the spaceship for fear of blowing fuses in the unsophisticated terrestrial mind with their talk of anti-gravity drives and the mysterious 'Elysium particle'.

The book was acclaimed by, among others, Larry Niven, who said of it: '*Dragon's Egg* is superb; I couldn't have written it: it requires too much real physics.' Arthur C. Clarke called it 'a knockout'. *Egg* was fresh, speculative, brilliant, verging on plausible. But what it shows, more than anything, is that there is no necessary incommensurability between science and literature, that 'science fiction' is not automatically an oxymoron. Forward said he owed it all to two things. First, the time that he went off to Europe as a Boy Scout for the Sixth World Scout Jamboree held in Paris and came back home and had to give public lectures, aged

15, on 'Whither Europe Now?' to women's institutes and rotary clubs. And second, the time his mother completed a jingle, 'It has more value in it . . .' by adding the words 'than at first meets the eye', thus winning several thousand dollars and paying for young Robert's college education.

His mother's jingle, originally associated with the manufacturing of shirts, may have linked itself in Forward's mind with the cosmos at large: maybe the universe, just like a shirt, has more value in it than at first met the eye, or telescopes or radio astronomy either. And so he was inspired to seek a non-optical, non-electromagnetic approach to understanding. But he was also impressed by the rate of $1,000 per word (which he said he never equalled). Thus he went on to become the 'guy that hard science-fiction writers go to when they want to know about the science'. And he kept on writing: science fiction (*Starquake*, *Marooned on Eden*, *Timemaster*, *Camelot 30K*) and science fact (*Indistinguishable from Magic*). But he never gave up pure physics either. He was a visiting professor at the International Space University. He contributed an article on interstellar travel to a Columbus 500th-anniversary book, published by the Smithsonian, entitled 'Where Next Columbus?' In 1994 he was the keynote speaker at the Practical Robotic Interstellar Flight Conference. He had the perfect name for the job and he used it to good effect (omitting the Lull part) when he founded a company, Forward Unlimited, which came up with the idea of 'space tethers'. The tethers, at their simplest, could reel in dead or dying satellites and thus stop junk accumulating around the planet; at their most complex and spectacular, they were capable of propelling ships along from Earth to Mars without all the expense and bother of rocket engines. They became 'space elevators' in Arthur C. Clarke's *The Fountains of Paradise*, but Forward was working on them for real (why should fiction have all the best ideas?) and NASA and the Department of Defense were rumoured to be giving them serious consideration.

He was still campaigning for laser-driven lightsails and antimatter-annihilation propulsion systems, and pondering the possibilities of time travel, when he died of multiple brain tumours in 2002 (his main regret being that an antimatter gun for shooting down brain tumours was still only on the drawing board). In the same year he was even giving papers at the Aeronautical and Astrophysical Association annual conference. He still had the original great quiff of hair, now white. He wrote his own obituary. It started: 'The intelligent pattern of protoplasm that had been Robert L. Forward ceased coherent operation on September 21, 2002.' But he can still be viewed in the form of pure light by extremely small people on a neutron star many light years from Earth, or read in the form of words practically anywhere by anyone.

In 1978 Forward published the paper that would revolutionize the gravitational wave field. Thomas Kuhn, in his *Structure of Scientific Revolutions*, depicted science as a broadly conservative institution that regularly coagulates into 'paradigms', great suffocating bundles of belief, textbook questions and answers, and periodically is shaken to its core and radically rearranged by occasional 'revolutions'. Einstein was one such revolutionary thinker. Forward, I would say, is another. One section of his autobiographical notes is headed: 'How I took on the world-wide established particle-physics community'. But then I think most scientists are – at least potentially – rebels who tend to look askance at the status quo and the establishment and are forever wondering if they could come up with something better that would blow everyone else clean away.

'Wideband Laser-Interferometer Gravitational-Radiation Experiment' (*Physical Review*, D17, pp. 379–90) showed for the first time that it was possible to build and operate a 'laser interferometer gravitational radiation antenna'. The trouble with Weber's resonant mass was that it was too short and too thick. It needed to be longer to be more sensitive. All that metal was just too heavy and awkward and cumbersome. How could you catch something as subtle and transparent as a ghost with something as clunky as a steam engine? At this time, lasers were becoming more available. I remember Unc showing one off to me in Southampton around 1970. Swinging back inevitably towards his mentor's first love, Forward was drawn to the highly coherent light form as early as 1961. And he also started to wonder what he could do with it. He first thought of riding a solar-generated laser beam all the way to the stars ('Ground-based Lasers for Propulsion in Space', 1961, and 'Pluto – Gateway to the Stars', 1962). It was no problem getting going, the only remaining difficulty, so far as he could see, was how to stop. Then, a few years later, it occurred to him that the laser could also be the solution to the problems of the Weber bar.

Essentially, Forward put the bar and the laser together, replacing all the aluminium by jets of light. Forward imagined a pair of beams, going off at right angles to one another, and then coming back again. More like a stethoscope than a telescope. The laser beams would be alternately tugged out and elongated and then pushed back in again and compressed by passing waves, like a concertina, or a bodybuilder's abdominal muscles in a series of crunches, alternately stretched and squeezed (engineers talk of 'strain'). Forward's pioneering device was designed to measure the fractional change in length between two end masses. All you had to do was register and measure the degree of interference in order to be able to 'read' the waves, the pulse of the patient. In one version, Forward conjured up a central axle with the laser beams and their terminal masses being spun around at high speed like an insanely speeded-up roundabout in a playground or a spinning top. But he worked out that you didn't really need all the rotation. Sheer length would be enough: the longer the arms and the greater the laser power the more sensitive it would be. By 1972,

Forward had already attained a level of sensitivity that would not be bettered for a decade.

Ironically, the laser, which Weber had more or less invented (without receiving the proper credit) and then shied away from, had been brought in to improve on and illuminate Weber. Forward's interferometer was a brilliant synthesis of intellectual history, bringing together the realm of pure light and the electromagnetic spectrum with the alternative path of grav waves. In the end, Forward had seen, they were all wave forms, and one wave could impact on another wave to generate bands of pure insight.

Forward set up the Interstellar Research Foundation as long ago as 1961. Even then, before Saturn V and the Apollo missions, he was already looking well beyond the rocket and the solar system, towards atom-bomb propulsion, ramjets, fusion drives, microwave sails ('Round Trip Interstellar Travel Using Laser-Pushed Light Sails', 1984), but he knew it was going to take time to come up with the optimal vehicle. With the gravitational wave detector, it was like he was already travelling effortlessly and without getting out of his chair, riding laser lightsails to the furthest corners of the universe. If we couldn't go to the stars (yet), maybe we could get the stars to come to us. LIGO could provide a rough guide to interstellar highways and byways to serve the future explorers who would eventually betake themselves to the far side of for ever. You could call it, as technicians did, a 'quantum nondemolition sensor'. But Forward had built a stairway to heaven, made out of steps of pure light.

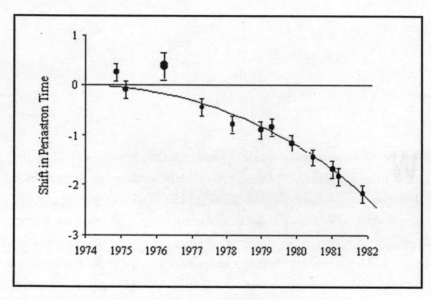

The Hulse–Taylor binary pulsar: the decrease in orbital period

19

'We're like Galileo,' said Dale Ingram. 'The day *before* he looked through the tube.' Dale was something like 'Information Executive' of LIGO. He had turned up, slim, bespectacled, smartly dressed, buttoned-up, in a suit and tie, while I was gazing at the Weber bar and working out how to stick it in my pocket and take it home with me.

'Only three of those still in existence,' he told me, proudly. 'Harvard have one, Caltech another. And we have this one.'

Then Fred Raab came out of the back room. He made a striking contrast to Dale. He was untidy, almost scruffy, open-necked shirt, heavily built, with a bristling moustache. He was the director of LIGO. Fred (partly because he always signed himself Fred, and also because I wasn't sure how to pronounce Raab, I called him Fred) was just about to leave. As a result of my ramble in the desert, the sun had gone down on our rendezvous and planned conversation. I knew he was due to go into hospital the next day for a mouth operation, minor but serious enough to render him incapable of speech. I had flown all the way up the

West Coast and driven several hundred miles to speak to him and now I had missed the boat. He didn't actually say 'You shoulda been here yesterday,' as surfers had a habit of doing, but it was firmly implied.

'What was it you were interested in again?' he said, apologizing for having to get going.

'Oh – everything,' I said, holding my hands out like a fisherman trying to indicate the size of the one that got away. 'Pretty much.'

Fred hadn't quite put on his hat and coat, but he had picked up his case and was moving inexorably towards the door. Obviously I had blown it.

'Doesn't it feel a bit like *Waiting for Godot*?' I said, following him. 'Or *Hamlet* without the prince?'

Fred stopped in his tracks. He was a big man. He knew what I meant. He put the case down again and turned on me. 'There's no question about the existence of gravitational waves,' he said, with some measure of indignation and righteousness in his voice, and there was a redness in his cheeks that hadn't been there before. 'They are real. The physics is clear. Especially after Hulse–Taylor. That reassured everyone. We knew they weren't just constructs. They had the energy. Whether waves exist is not up for debate.'

Fred snapped a finger at a large framed graph on the wall. No pictures of cottages with smoke coming out of the roof or waterfalls or sylvan gardens or sunflowers in this lab. The caption below read 'PSR B1913+16. QED'. Twenty years before in Puerto Rico, Hulse and Taylor had discovered a binary pulsar, two neutron stars spinning like whirling dervishes and dancing ever closer to one another 16,000 light years away with an energy depletion that was only explained by the emission of gravitational waves (hence QED). The 'inspiral' (so-called because the two stars were spiralling inward towards one another) matched the theory perfectly, to within a third of a per cent. Fred had it pinned up on the wall like a gorgeous butterfly. I stood in front of this reassuring picture. It was clear we were not on some wild-goose chase. But I had a feeling there was a 'but' coming.

'But,' Fred said, 'can you build a detector that is sensitive enough and run it continually enough to pick them up?'

'I don't know. Can you?' I said.

'It's hard, but the physics is OK. It has to be continuous,' he said. 'The waves don't wait for you to do it one weekend in the month. You have to be able to accumulate signals for long enough.'

'I heard you picked up a lot of trucks going by and rabbits chewing and . . .'

'You have to be able to get above the noise floor,' said Fred, cutting me off. 'The challenge is to get the floor down so there is an adequate opportunity to make the detection.'

Unlike the Keck, it wasn't just earthquakes that LIGO was worried about: it was like every falling leaf was an earthquake, from their point of view.

Fred Raab was one of the four authors of the original proposal for LIGO. He had started off in precision experimental physics and molecular spectroscopy. He was famous for measuring very small things, totally imperceptible, from the great realm of the unseeable: stuff like the width of a particle or the mass of a beam of light, that sort of thing, or using atomic clocks to time how long it takes a hummingbird to flap its wing. He became obsessed, enchanted, with measuring everything, especially where it deviated from perfect symmetry. If you looked closely enough, everything was asymmetrical. The world was irregular, grainy, textured, fractured. All that stuff you thought was smooth – paper, ice, glass, silk: think again, turns out it's as bumpy, lumpy, and potholed as a farmyard track. He was at Washington State, on a post-doc, when he got the call from Kip Thorne (which is a little like an actor getting a call from Steven Spielberg), who asked him if he wanted to get into LIGO. It was the ultimate challenge for somebody who loved measuring. No one had ever really been able to measure a wave before. It was funny that cosmology, in a way the science of the extremely large, dealing with planets, stars, constellations, galaxies, the whole damn thing, should, in the end, boil down to something so small. How could he resist that?

'It's a bit like looking into the void,' he said (in fact, I thought, that was exactly what he was doing, it wasn't an exotic metaphor). 'But you ask yourself: is it likely that sooner or later someone will succeed? I thought that it was. I wanted to be there when it happened. I didn't want to read about it in the *New York Times*.'

LIGO was having an upgrade when I came to call. It was always having an upgrade but this was a big one, known as 'Advanced LIGO'. 'We know,' said Fred, 'we can implement more powerful lasers, better mirrors, better signal processing, a better vibration solution.' Something called a 'signal-recycling mirror' was being added on. At the end of the upgrade they would be able to scan as much in an afternoon as they had been doing in a year. The chances of picking up a wave increased proportionately, from long shot to possible and even probable. They were expanding the field from 10 kiloparsecs to 10 megaparsecs – a billion times as much volume.

Everything takes time. They got the original grant in 1992. They were building from '96 to '98. Installation of instruments: 2000. The first engineering run, 2002, then the service runs. By 2005 they had attained the sensitivity levels that LIGO had been designed to meet. And they had forced the costs down. Now they were trying to get ahead of their own targets. They were in the midst of an all-sky sweep that would run up until September (instrument enhancements permitting). Some said the LIGO guys were 'arrogant', but they had run into so much scepticism and flak they needed to be just to keep going. It was like intellectual armour-plating.

Fred saw me glancing admiringly (and covetously) at the Weber bar. 'Weber was the pioneer. He was our Edison. He convinced the community it was doable. It took a lot of imagination and guts.' Fred picked up his case again. 'Many things that looked impossibly hard when they were first done are now standard and nobody thinks twice about it. Cellphones, for example. That's how it will be with grav waves. They'll become commonplace.'

I realized who it was Fred reminded me of. It was Maigret, the great detective. Or maybe his NYPD counterpart, since Fred was originally from New York, born and raised in Queens (the tough-guy borough): large, gruff, grouchy, baggy, Kojak with hair, probably eating too many doughnuts, used to getting shot at, and unstoppable. A grizzly bear of a man. Detecting had become his life. Especially the hard cases, anyone could round up the usual suspects. It was like he had been pursuing the perpetrator for years and he had kept on getting away but Fred was never going to give up. He was relentless. In the end the guilty party would be detected, apprehended, and then put away, for ever, on the hard drive. The lab, he reckoned, was the Supreme Court, in which all the evidence could be tested and felons finally sent down and innocents set free.

He sounded like a crusty old Supreme Court judge who has seen everything when I ventured to ask him about string theory, the idea (strongly advocated by some physicists, notably Michio Kaku and Brian Greene) that all matter in the universe consists of extremely small pieces of string-like stuff vibrating at high speed, quite possibly in eleven dimensions, all folded up, very tight, like some kind of microscopic origami. Strings had a lot of smart fast-talking mathematicians to defend them. Fred kicked them all out of court. They – he reckoned – had absolutely no witnesses and no

Exhibit Number One. It was all tell and no show, like some kind of ancient prophecy or lyric poem, only with more in the way of equations. It was evident that Fred would rather have a lot more of the show and a lot less tell. There was an idea that the early 'cosmic strings', larger versions of the smaller lengths, could still be around (after all, according to the conservation of information, if they *had* been around they were *still* around) and you might get them emitting waves.

'I was *surprised*,' Fred said, with heavy irony, 'to hear that we were supposed to come up with evidence to support string theory'. The year before he was at the Santa Barbara Institute for Theoretical Physics when a bunch of theoreticians told him they were relying on him to save the strings. But he didn't want to be anyone's saviour. Fred was an experimentalist and a measurer who was sceptical about pure theory and theoreticians. 'It's more religion or poetry than science,' he said. 'It's beautiful – but is it true?' He wrapped it all up and rejected it in a single acerbic epithet: *unfalsifiable*, the word the philosopher Karl Popper had used to put down psychoanalysis and Marxism. 'Whatever evidence we come up with, they'll juggle it around to fit. That's the trouble with string theory – there's nothing you can come up with to prove it's wrong. It has infinite wiggle room.'

Fred Raab had shifted away from his youthful obsession with anarchic asymmetry. Chaos, anarchy: it was fine in exceptional cases, but it all came down to laws in the end, universal physical laws, and he was determined to make sure that everything conformed, was consistent in a strict Newtonian way. Like Rømer, he didn't want any deviants and scofflaws running around causing trouble. The universe began, to his way of thinking, when it decided to become law-abiding: 'The beginning,' Fred argued, 'is the point at which we have confidence that physical laws will work.'

It seemed slightly paradoxical therefore that he was so concerned with sensitivity. He did not look like a particularly sensitive kind of guy. He was tough as old boots. He was hard, he even liked the word 'hard'. 'When I got into this business,' he growled, 'we all knew it was hard.' Yet he spent all his working days trying to 'double the sensitivity', sounding for all the world like an ad for a male contraceptive, and wondering fussily whether you could 'get the sensitivity up'. That's what the signal-recycling mirror was for, and all the upgrades, and enhanced resonance. 'If you can double the sensitivity,' he said, 'you can see twice as far, and eight times the volume – and eight times the data. That's where we score over the old telescopes.'

Fred was a practical man. But there turned out to be a deeply philosophical side to him. In a way he had to be philosophical, he had no option, it was thrust upon him. Let's face it, so far he had absolutely nothing to show for the years of effort and investment. No wave. Nada. Our nada who art in heaven. He pushed open the door and walked out. I watched him go. I saw him throw in his case and get into his car. He had things to do, places to go, people to see, an operation to prepare for. I stood at the front door as if I was the director of the whole lab and he had been interviewing me. Then I saw him get out of the car again and walk back. He wanted to ask me a question.

'You want to know the big difference we've already achieved?'

'Yes,' I said.

'We've changed the way people think. And that is a big change. Look at us. What do you see?'

I stopped to take a look at both of us.

'A couple of guys talking to each other.'

'Exactly. You are using your senses to detect an interaction between atoms.'

The way he said it, I could tell that Fred didn't think all this 'using your senses' was such a great idea. But he was right of course. And he had put it rather well, I thought: we *were* a bunch of atoms interacting. 'We are electromagnetic beings. We experience things electromagnetically. We're made out of light. So naturally we *look* at ourselves and we *see* things. But did you ever think that could be a mistake?'

'Plenty of times,' I said. 'In fact just about all the time, come to think of it.'

'Do you think you can *see* a star?'

I paused to reflect. 'Well, I used to think I could, until about two seconds ago, when you asked me that question.'

'What you are seeing is your inability to resolve the image. Even if you take the largest optical telescope.'

'Like the Keck?'

'Whatever.' Fred didn't have too high an opinion of the old-style optical telescopes. Extremely Large, Ultra-Large, Adaptive Optics, it was all the same to Fred. He thought of traditional star-gazers as roughly on a par with moths, fluttering automatically towards the light. 'The stars are unresolved. They've never been resolved. They're just point sources of light. Everything else is just distortion. You haven't in fact seen a star.'

Twinkle, twinkle, little star. We love the twinkle. We are so transfixed by the twinkle that we sing about it to our children in the cradle. 'Like a diamond in the sky' – wasn't that some kind of drug-induced hallucination? Whatever it was, it wasn't science. I thought of the red dot I'd seen at the Keck, 13 billion light years away. That dot was truer than all the more spectacular twinkling stuff. On the one hand was everything that exists (stars), on the other hand was everything you could see (twinkle), and there was a big mismatch between the two. Fred thought the pre-Copernican assumption that we are the centre of the universe was completely reasonable, because that is how it *looks*. But anyone preoccupied with mere *looks* was an idiot. Everything you could see was essentially a lie, an electromagnetic lie. And it was the same with time as with space: we are electromagnetically programmed to have a sense of time as flowing forwards: we don't like to watch a movie backwards.

'In EM [electromagnetic] astronomy, you're looking at *emitters*; with GW [gravitational waves], you're sampling different things. To have a totally independent way to bring in information from the universe,' said Fred, 'that would be a big deal. Most people just naturally assume if they see something that is the way it has to be. But it doesn't. We've learned a lot doing things that way. We have two-megapixel images of the early universe. They're wonderful and they're all over the WMAP website. But what if there were other wavelengths out there, ones we were oblivious of, like whale song? Like whole dimensions that are invisible to us?'

'Well,' I said, giving the question serious consideration. 'Then we would be none the wiser, because we're oblivious of it.' I couldn't help thinking of that old road sign in Hawaii: you could *be*ware invisible cows, but if they were invisible how could you ever be *a*ware?

'But we're not oblivious any more. We know grav waves are out there. But we have to retrain ourselves to detect them. We're not used to it. We still have the old electromagnetic bias. But ultimately we'll come around and it will be like a whole new sense that we haven't even discovered yet. A real sixth sense. Tuned in to waves. And for that we need a different probe.'

'So that we can feel the vibrations, you mean?' I had the old Beach Boys song playing in the back of my mind, something to do with 'picking up good vibrations', which rhymed with 'she's giving me excitations'.

'We can't see them, that is for sure. Maybe we won't be able to feel them either. We don't even have the right vocabulary for what we are doing. But that is why we sometimes talk about "hearing" waves: to get around "seeing" for a change. And because they are more like sound waves than the standard electromagnetic variety.'

Fred was offering me nothing less than a vision (except 'vision' was obviously still the old-school vocabulary that needed to be overhauled) of things to come. A new epistemology. He certainly wanted to check out the inside of black holes and see, just as I did, all the way back to the beginning of time. But there was more to it than that. 'You're talking about changing human beings fundamentally then?'

He looked at me like the grizzled, veteran NYPD detective again, sizing up a suspect. 'And that would be a bad thing?'

I'd often wondered what it was that Nietzsche meant by his enigmatic remark 'What if truth were a woman – what then?' Quite unexpectedly, Fred Raab had invested it with some kind of sense: maybe we didn't exactly think that truth was a man either, but all along we had been thinking what we saw must be truth, the whole truth and nothing but the truth, the old electromagnetic bias, in short; whereas it could well be that truth was something stranger and richer and less self-evident and more wavelike than we had ever imagined. It was Hamlet's Law, There are more things in heaven and Earth than are dreamt of in your philosophy. And Fred, like Hamlet, was having difficulty with a woman, with the truth. Despite truly heroic expenditure of effort, the great interferometer had so far remained cool and resolutely unexcited, failing to respond to his touch. It was a big shift, a change of gender virtually, from the old telescopes. With the scopes all you had to do was make them big and poke them up in the air and they would invariably come up with something; with the interferometer, a nexus of tubes, you had to wait, patiently, maybe forever, until there was a very slight sigh and a shiver. We had moved on from the simple preoccupation with sheer size and power to an emphasis on sensitivity. It was no coincidence that the next generation of sensor was going to be called 'LISA'.

Fred stalked off again into the dying light. The head of LIGO was a man on a mission. He liked to concentrate on the engineering, and fixing things, but when he was pushed to the limit, it was clear that he had a secret moral and metaphysical agenda. This is not a quote, and I don't want to put words in his mouth, but I had the impression that he felt that gravitational waves – always assuming we could actually pinpoint them – were coming to save us: to burn out the old ways and install some radically new software in our brains. And until we did finally get our heads around gravitational waves, we were doomed to carry on in the same old ways, ways that would probably lead to our destruction and certainly to our perpetual confusion.

Fred Raab had left me, seemingly, in possession of the entire laboratory. The whole place had gone eerily quiet again now he had gone. No Fred, no Dale. I was left kneeling reverentially in front of the Weber bar, shyly caressing its sheer bulk and smoothness, and thinking. One thing Fred had said, in his philosophical mode, stuck in my mind. 'The slaves in Plato's Cave, they thought they were seeing the real thing too.' He was referring to the classic passage in Plato's *Republic*, his dialogue on politics and truth, where Socrates conjures up the image of a cave as an allegory of our experience on Earth. We are prisoners, chained up in the cave. All we can see of people going by and the whole outside world is shadows projected on the inner wall. Our mistake is to assume that these shadows are everything that exists: that this is what reality is like, or take out the 'like', there is no difference between seeming and reality. And then one day one of the cave-dwellers escapes his chains and gets out into the real world and is amazed to discover there is something other than shadows. He returns to the cave – joyfully – to report his discovery to his comrades who are still inside. He thinks they are going to be pleased with the news. Instead, they think that he is either mad or a radical subversive of some kind, and kill him, and go back to the shadows. Something like it happened to Socrates, who was put to death for corrupting youth with his inflammatory ideas. Fred Raab seemed to be staking a claim for being the new Socrates with the wave as the fundamental Form (or Idea in Platonic theory) of reality.

'The question is,' he said, 'when we look at the universe: is this the real thing? Or are we just limited, am I in chains?' His idea was that grav waves, just as they passed right through solid matter, would enable us to see right through all the old illusions – to see through seeing. The only problem was we had to see them first, or hear them, or something.

There was one more of Fred Raab's allegories that haunted me. He had brought in a story about a spider. I think he was using the spider as an example of another limited point of view. 'When he feels vibrations through his feet, he finds that an insect – or lunch – has turned up.' It was a reliable enough method with proven results, but the spider's universe was 'too thin', and we in a similar way are restricted by our own particular way of seeing things and chopping them up into bits or wrapping them up in our own silken webs and sticking them in the larder we call knowledge. On the other hand, turning things around, I couldn't help but wonder if Fred suspected that the spider, in his own way, with his highly sensitive hairy legs, was on to something and that feeling vibrations coming up through your feet was just as reasonable as any other mode of perception, maybe more so. And that LIGO was just that: another mode of perception, wired up to detect vibrations instead of light, less a telescope than an intricate spider's web poised to collect the waves that would ultimately flutter into it. It would reveal 'the thing inside the supernova'; it would take you back not just to the early universe (like George Smoot and WMAP) but all the way back to 'time zero'. The Beach Boys with their good vibrations, teenagers jiving at the high-school hop, hula dancers, a guy with a pair of bongos between his knees, and spiders dancing on their webs: all those, in short, who could tune in to the rhythm of life, feeling it come up through the floor with their feet, were – without knowing it – highly advanced students, potentially, of the structure of the universe.

I stood out on the porch as Fred drove out of the car park. He slowed down and wound down the window. 'If you're going to be a searcher,' he said, pulling away in a cloud of dust, 'why stop? You want to keep going.'

20

Through the windows you could see the wide voluptuous curves of the Columbia River snaking through the trees. I was pedalling like a maniac, going nowhere. John was lying down and pushing himself to and fro with his legs. We were both gleaming with sweat. We were staying in the Shilo Inn in Richland at the time and we had chosen the same moment to go and work out in the gym. There was an eccentric system to do with handing in your room key at the desk and receiving in exchange a key that would open the door of the fitness room. And John and I had had to pass the key around so we got to talking.

It turned out that he had driven over from Seattle just as I had driven from Portland and we compared times and I ranted about the elusiveness and remoteness of LIGO. John smiled benevolently, even though rowing at the time. He was now selling scientific equipment, but he used to work in one of the many labs around Richland. The Hanford complex (of which LIGO was one part) employed a lot of scientists. 'This city has the highest number of PhDs per head of population of anywhere in the

United States,' John said. He had read a survey on the subject in some respected journal.

'How long were you here for?' I said.

'Couple of years,' he said. 'Then I had to get out.'

'Too much fun?'

'It's like doing a PhD thesis. But for ever. You go mad in the end. I wanted to leave while I was still holding on to some shreds of sanity. And I had a better job offer too.'

I stared at the window, watching the wide, languid river roll by, wondering where it came from (mountains somewhere to the north, I guessed) and where it was going.

'What do you think of the electromagnetic bias?' I said.

'I think you've been talking to some pretty strange people,' he said.

He almost certainly had a point there. To John – and to almost everyone else of an electromagnetic persuasion likewise – LIGO land was like some kind of lunatic asylum, populated by schizos and neurotics obsessed by 'sensitivity'.

While we were talking, a whole family trooped in, man and wife and their two children. Without wanting to sound like a total body fascist, I think it would be fair to say that they were not in supremely optimal shape. Flabby, Puffy, Flaccid and Pouchy. In a Disney cartoon they would probably have names along these lines. But at least they had come to the fitness room, I observed inwardly. Good for them. Working on perfecting the body. In fact they spent the whole time lounging around the jacuzzi and taking no perceptible exercise at all and discussing competing brands of potato chip, loudly. Partly because I had an idea of making use of the jacuzzi myself and they had completely colonized it, I found myself resenting them, as one does.

And then I realized that I was still guilty of my old electro-magnetic bias. I couldn't seem to get over it. To my eye, they all looked overweight. Of course I was applying a standard of anatomical perfectionism and finding them wanting, or, in a way, not wanting enough. But beneath all that I was applying an even more fundamental cognitive grid, I was relying on my eye and my ear, and judging accordingly. But if only I could overcome my factory settings (so to speak) and perceive these entities, soaking in the baths, as some kind of pure wave form, if I could only attune myself to pick up their good vibrations, perhaps I would understand that they were already perfect.

There is a story (which I read as a child in an issue of *Astounding Tales* or *Fantastic Adventures*) of a tremendously kind but extremely ugly man who gets on a bus. He is so ugly that everyone else shies away from him and eventually they all get off. The man is depressed that everyone treats him with revulsion when he feels nothing but love for his fellow beings. The bus keeps on going and soon it is taking off and flying away through the clouds at enormous speed. The man was only planning on going into the middle of town to buy some groceries so he is surprised when he looks through the window and see stars. The bus lands on another planet where the man is welcomed with open arms by extremely beautiful people. 'But you're not running away from me,' the man objects. 'Can't you *see* how ugly I am?' He has become used to people flinching and averting their gaze. 'To us you seem wondrous,' says a particularly radiant woman. 'On this planet we see only the soul. What is on the inside not on the outside.' They had spotted the man's tremendous soul from far far away, shimmering out at them, and arranged the magic bus to bring him to them.

That, in a way, was what LIGO was after: the soul of the universe. It was only after I came out of the Shilo gym that I understood that this was exactly what I was pursuing too.

21

I sat on a bench by the great river that evening, watching the moon reflected in the water. I had taken a book of short stories with me. One of them was by Jorge Luis Borges, the Argentine writer, and recounted a quest to discover the impossible and irresistible Aleph, the one thing that contained everything, infinite space bounded in a nutshell. It struck me, reading the story, occasionally looking up at the insects driven mad by moonlight, that Borges was, in a way, the Fred Raab or Joe Weber of literature. When his hero (who is also the narrator) finally tracks down this magical object (that contains all objects), on the nineteenth step of a dusty cellar in Buenos Aires, he perceives a small iridescent sphere of almost unbearable brilliance. He realizes that, although its diameter is probably little more than an inch, all space is there, the entirety of the cosmos, actual and undiminished. What he sees, as if gazing into a crystal ball, but from every conceivable angle, is recorded and celebrated with a sentence that is probably one of the longest and the most perfect ever written (and therefore impossible to condense).

I saw the populous sea, I saw daybreak and nightfall, I saw the multitudes of America, I saw a silvery cobweb in the centre of a black pyramid, I saw a broken labyrinth (it was London), I saw, close up, unending eyes watching themselves in me as in a mirror, I saw all the mirrors on earth and none of them reflected me, I saw in a backyard of Soler Street the same tiles that thirty years before I'd seen in the entrance of a house in Fray Bentos, I saw bunches of grapes, snow, tobacco, lodes of metal, steam, I saw convex equatorial deserts and each one of their grains of sand, I saw a woman in Inverness whom I shall never forget, I saw her tangled hair, her tall figure, I saw the cancer in her breast, I saw a ring of baked mud in a sidewalk, where before there had been a tree, I saw a summer house in Adrogué and a copy of the first English translation of Pliny – Philemon Holland's – and all at the same time saw each letter on each page (as a boy, I used to marvel that the letters in a closed book did not get scrambled and lost overnight), I saw a sunset in Querétaro that seemed to reflect the colour of a rose in Bengal, I saw my bedroom and no one was in it, I saw in a study in Alkmaar a terrestrial globe between two mirrors that multiplied it endlessly, I saw horses with flowing manes on a shore of the Caspian Sea at dawn, I saw the delicate bone structure of a hand, I saw the survivors of a battle sending out picture postcards, I saw in a shop window in Mirzapur a pack of Spanish playing cards, I saw the slanting shadows of ferns on a greenhouse floor, I saw tigers, pistons, bison, tides, and armies, I saw all the ants on the planet, I saw a Persian astrolabe, I saw in the desk drawer (and the handwriting made me tremble) unbelievable, obscene, detailed letters, which Beatriz had written to Carlos Argentino, I saw a monument I worshipped in the Chacarita cemetery, I saw the atrocious remains of what had once deliciously been Beatriz Viterbo, I saw the circulation of my own dark blood, I saw the mechanics of love and the modification of death, I saw the Aleph from every point and angle, and in the

Aleph I saw the earth and in the earth the Aleph and in the Aleph the earth, I saw my face and my viscera, I saw your face, and I felt dizzy and wept, for my eyes had seen that secret and conjectured object whose name is invoked by all men but which no man has looked upon: the inconceivable universe.

This circuitous, manifold, magnificent sentence, more than 400 words long, swelling and peaking like a wave, contains so much, more perhaps than any sentence ever written before or since. And yet it leaves out nearly everything and no one was more acutely aware of this than Borges himself. 'Any listing of an infinite series is doomed to be infinitesimal', he wrote. In a kind of self-defensive gesture, Borges spoke of an 'aesthetics of omission' and wrote only the shortest of short stories to display the very inadequacy of which he was conscious: the shortness, the inescapable brevity and exclusion of each and every text. 'The Aleph' surveys Scotland as well as Argentina, time as well as space; all the mirrors, all the ants. But the only planet to get into it is the Earth. Not even the moon is mentioned. We are more confined here than in almost any science-fiction novel. The timeframe is barely more accommodating (we see one woman alive and then dead and decomposing, encompassing less than a century). Fred Raab, reading this story, would surely have pointed out that Borges, with his exhausting repetition of the phrase 'I saw', was absurdly monopolized by the electromagnetic spectrum (not much in the way of smell, or hearing, for example.) Even if he really could see everything he would still be fabulously limited in his perception. Unc reckoned there were approximately 10^{123} bits of information in the universe, so really Borges didn't stand a chance of getting it all down in a short story, not even a long one.

And yet the desire to reassemble all the disparate parts of a lost whole is manifest. Borges is capable of imagining mirrors that *do not reflect him*, a world from which his own subjectivity has been expunged. His point of view has become the point of view of the cosmos. His conclusion remains a melancholy one: that the universe is 'inconceivable', it is 'that secret and conjectured object whose name is invoked by all men but which no man has looked upon'. Like Galileo, he does not claim to have discovered 'how His hands built it'. Like Galileo at the end of his life, Borges – when I met him anyway – was blind. He had a pretty young Japanese woman to guide him around and act as his secretary. He had, to some extent, transcended the electromagnetic. It was impossible for him to look upon anything. And now he existed (he exists) only in the form of light and words.

I closed the book. It was just a slim volume and it looked even slimmer when you set it against the broad expanse of the river, vanishing into the darkness, with no beginning and no end. The water, like a mirror, caught the light from the moon (which was another mirror) and reflected it at odd, rippling angles. A reflection of a reflection: that was all literature (and art) amounted to. Most writers just try to get the minute bit of the universe that they can actually detect fairly straight (realists). Others, who can see the absurdity of the attempt, without being able to see beyond it, admit their own limitations, everyone's limitations, and try to tear up the part they can see into even tinier ever more meaningless shreds (surrealists, deconstructionists). But there are a few brave souls who try to reconstruct the whole, to revisit the original singularity, to take a few crumbs and somehow fashion them into entire loaves and fishes.

Victor Hugo, for one, in the middle of the nineteenth century, revolted against the positivist accumulation of detail. To write about nothing was impossible: 'There is no such thing as nothingness,' he wrote in *Les Misérables*. 'Zero does not exist. Everything is something. Nothing is nothing.' At the same time, it would never be enough for him to write merely about *something*. He fantasized about writing 'the complete book' that would somehow contain *everything*, like an encyclopedia, the ultimate *Schoolboy's Book of Knowledge*. His epic poem *The Legend of the Centuries* comes close. He invented heroes who, being immortal, would eventually get around to witnessing everything that happens, the past and the future. He imagined aliens meeting on faraway planets, 'look[ing] at one another in the shadow, monster gazing at monster'. In exile on the island of Jersey, having fallen out of favour with Napoleon III, Hugo attempted to bestride life and death and communicate with the beyond via his *tables tournantes*. He had conversations with Shakespeare (who admired Hugo's work), Dante, Plato, Galileo, Robespierre, Beethoven, and Jesus Christ, but he was a little sceptical about 'Mozart', whom he suspected of just pretending to be Mozart. Other less elevated souls had already been reincarnated as a flower, a scaffold, algae, a meat-hook, a cradle (Herod), and phlegm (Judas). Notre Dame cathedral struck him as the kind of place that seemed to encapsulate history, a second Tower of Babel: each face, each stone seemed to him to represent 'not just a history of its country, but of art and science too.' An angel, a spirit, Napoleon, a gargoyle: each of them is an 'abbreviation of the world', a microcosm, an Aleph.

Borges and Hugo each put me in mind of my boyhood favourite, Jules Verne. They all seemed inhabited by the same spirit, driven by the same relentless quest as I was. Verne span his heroes around from one pole to another or swung them around the equator in eighty days flat and then drilled them down into the centre of the Earth, only to raise them up again and fling them out at the moon. If they had no rocket, they would hitch a ride on a passing comet or take a submarine 20,000 leagues under the sea. To Verne's way of thinking, there was no such thing as an aesthetics of omission, every omission was a failure. He hated to leave anything out. Every short story – every poem, however epic – was too short. Asked how much he had written, he once replied, 'About three yards.' I'm not sure if that was a boast or a lament. The preface to the *Voyages Extraordinaires* baldly proclaims the author's modest ambition 'to sum up all the knowledge amassed by modern science, geographical, geological, physical and astrophysical, and to rewrite the history of the universe'. Fifty or more books later, he felt there was still a long way to go, but like Fred Raab he was determined to keep on searching:

> It is my intention to complete, before my working days are done, a series which shall include, in story form, my whole survey of the world's surface and heavens, there are still left corners of the world to which my thoughts have not yet penetrated. As you know, I have dealt with the moon, but a great deal remains to be done, and if health and strength permit me, I hope to finish the task.

Verne died too soon to complete his heroic task. But in LIGO land the hunt for the Aleph was still going on. Borges says that we are all inadvertently chasing it every time we open our mouths: 'in the human languages there is no proposition that does not imply the entire universe; to say *the tiger* is to say the tigers that begot it, the deer and the turtles devoured by it, the grass on which the deer fed, the earth that was mother to the grass, the heaven that gave birth to the earth.'

I wasn't so much on some kind of lunatic personal quest, a road trip to truth, I was just following the river, flowing with the current, drifting in the general collective direction of all language and thought. When I said (to myself) 'I go LIGO,' it should really have been the first person plural. Not ego but 'we go'.

22

I sincerely hoped that I would, in the end, be able to relinquish all my old prejudices. But not just yet. I realized that so far I had only 'heard' LIGO, but I still had to 'see' it. The next day I went back there and we entered into some serious discussion, mainly about T-shirts.

I was in the control room. It was like the deck of the starship *Enterprise*, with dozens of people trooping about, fiddling with knobs and scrutinizing monitors. I was beginning to realize how many people there were on board this mission. The giant screens on the wall were like windows through which I was seeing the whole universe go by, at warp speed. In graph form. 'NS/NS inspiral range neutron star': one window opened out directly on to the Hulse–Taylor binary. The crew had to admit they might have a bit of difficulty translating it back into familiar electromagnetic terms. 'What are we going to put on the T-shirt?' – that was the question, the way they formulated it.

The astronomers at Keck world could strut their stuff with the electrifying image of the Red Square. The guys at WMAP control had the cover shot that had appeared around the world with the headline 'THE FACE OF GOD', of spotty, dotty lights bubbling up out of the plasma, like a pot of oatmeal coming to the boil, pointing the way towards galaxies yet to come. At the Large Hadron Collider at CERN in Geneva, they had the Higgs boson, the so-called 'God particle' (even if no one had actually seen one yet). But at LIGO they were so used to transcending the visual that they could hardly dream up an image to go on the merchandise. LIGO lacked a logo, that was the problem.

'How about a question mark?' said one guy.

'You want to admit you don't know?' The objector swivelled back to his workstation, intent on banishing ignorance.

'A big Pipeline wave,' said another. 'It'll look good anyway.'

'We can produce a space-time image,' said Evan, a confident, clean-cut third-year grad student, aged 24, from the University of Michigan. 'A GW map. Better than WMAP. We can go past that. Beyond photons.' He had been working at LIGO land on and off ever since he first visited the place as a kid on a summer internship and got sucked in by the insane grandeur of the whole project, taking a couple of beams of light and bouncing them back and forth like tennis balls, and gleaning information about the furthest reaches of the cosmos from the way they bounce. 'I thought, I can't pass this up. You're getting in on the ground floor of new ways of observing space.' Now he'd been there for seven summers. 'We're on the cusp of seeing real things. But we are going to be seeing things that are very different, maybe so strange people won't even understand them. The curves and warps and kinks of space. Whole geometries rather than simple objects. The system as a whole rather than the parts. Bulk motion. It's a different mode of perception.' This was exactly what I wanted. I already knew about 'objects'. But what I didn't know about, what nobody knew about, was what the philosophers call 'the thing-in-itself', the *en-soi*, raw, unmediated stuff, uncluttered by our presuppositions. It would be like seeing God, or the God particle.

'Can you show me the origin of the universe?' I said. 'Can you put that on the T-shirt?' Inadvertently, I had let off another storm. Whistles, sighs, the sharp intake of breath.

'That's like asking to see the Wizard of Oz, man. Nobody sees the Wizard.'

'The stochastic wave background.'

'The Big Bang?'

'The only things we are going to see right now,' said Evan, 'have to be close enough and violent enough.' If they were too close, we would simply cease to exist. Too far away and they might not even register. LIGO needed to calibrate itself on some medium-distance compromise catastrophe (like a star exploding or a neutron star or a black hole feasting on its neighbours) before it could home in on the origin. The bigger the event the better the wave.

'So, in reality, you have nothing to show?'

'If we see something, we would consider ourselves lucky,' Evan admitted. He wasn't too worried about it. He didn't have to discover anything to write his PhD thesis: he was concentrating on methods of calibration. He was still excited by calibration. 'We've proved we can build sensitive enough instruments. These are the most sensitive instruments known to man.' It was like actually discovering something with them would be an added bonus.

But I had learned one thing from Evan about mirrors. Every time light hits a mirror, it pushes on the mirror, and the mirror actually recoils. It vibrates, it pulsates. Light exerts pressure on a mirror, very slight, but measurable. Almost like a trampoline with photons bouncing up and down on it. I had often thought of how a mirror distorts the truth, reversing the object and showing its past rather than its present. Is this really me? But I had never considered before that every time I switch on the bathroom light and look in the mirror, I am basically wearing it out. Knocking out a few stray electrons. Eroding it, slowly but surely, like the sea ramming up against a cliff. The idea of a face so ugly it could break the mirror was only a slightly exaggerated caricature of what takes place every time we look in the glass, irrespective of our ugliness. And LIGO relied on that effect. All the old telescopes had something of the fairy tale about them: 'Mirror, mirror on the wall, who is the fairest of them all?' Whereas the interferometer turned it around and examined the mirror for signs of wear and tear, scars, wrinkles, distortions. I realized that no one had ever properly looked *at* mirrors before, we had only looked *through* them, seeing through a glass, darkly. But if you wanted the truth, the point was to inspect the glass itself.

23

Mike Landry ushered me out of the control room in search of a quiet place. Mike hated noise. He spoke fervently of 'beating down' noise, 'attacking' it, and 'lowering the noise platform'. He was Senior Scientific Officer and had day-to-day responsibility for work on LIGO. I asked him how much of his work was to do with eliminating noise. '*All* of our work,' he said, 'is to do with stamping out the resonances we don't want.' Maybe that explained why he was so mild-mannered and soft-spoken: he didn't want to speak too loud or get stressed out in case it upset the instrumentation. Mike was a Canadian and reminded me of Mulder in *The X-Files*. He was like the ultimate hi-fi fanatic. Extreme purity was his life work, his vocation, his dream. He pulled up jazzy images of the progression of the 'noise curve' which showed it dipping down inexorably towards (while never quite attaining) zero, or (as he put it) 'zero zero mode', the last word in zero tolerance. There was something at once benevolent and harassed about him, as of a young father, trying desperately to get his baby clean then wrap her in great thick layers of cotton wool and swaddling clothes and shut out the ghetto-blasting neighbours with a view to finally getting this beautiful but hypersensitive creature off to dreamland.

There was a hint, too, of the Grand Inquisitor about his mission, hunting down and eradicating any last faint residual heresies, anything that might interfere with or distort the voice of God, 'the pristine beam'. The underlying theory was that if you could only get rid of the impurities, then what remained must be the truth. Truth would just naturally and automatically present itself for inspection. All present and correct, sir. This was what Mike called 'the faith element'.

When he arrived here in 2000, fresh from the University of Vancouver, convinced that pinpointing gravitational waves was a real possibility, he made a disturbing discovery: 'We found that we were a very good detector of airplanes.' The planes would shake the air; the air would shake the optics. He fixed that, but that was just the beginning of the shakes. Mike was an old hand at following the traces of protons in accelerators. They seemed to have a width of 10^{-15} m. But now he was hunting even smaller prey, of the order of 10^{-18} m – one-thousandth the diameter of the average proton. Not so much a tall order as a small order.

Joe Weber had put his finger on it: 'Suppose you see a big pulse on one detector. You can't be sure whether that pulse was due to a garbage truck colliding with the building, a lightning strike, or student unrest.' (Students in the sixties had a habit of daubing the walls of his lab with radical graffiti.) The trouble with the Earth is that it is not very good at registering truth. It is such a noisy, feverish, querulous sort of place, shouting out its own local gossip, transmitting commands, prayers, reports, requests, love letters, parliamentary legislation, lies, songs, symphonies, match commentaries, statistics, acts of fraud and infamy, sobbing, old black-and-white movies, and endless radio chat shows, all being pumped out into the ether at all times of day and night. Not to mention all the journeys to and fro, and especially up and down, skipping, jumping, bouncing on pogo sticks, horses galloping, trucks walloping along the highway, trampolinists, tap dancers, holes-in-one, xylophones, people falling out of bed, people falling into bed, or jumping out of high windows. Me tapping on the keys of this keyboard. Everybody forever leaping and thumping. The whole planet is being constantly rattled around, it is never at peace. How can we ever hear anything but ourselves? God finally speaks and delivers the divine word and everything you ever wanted to know about everything and all we can hear is the sound of a toilet flushing. We are deafened by our own percussive stomping around ('anthropogenic noise'). And that is not counting earthquakes, tremors, seismic jolts, geological burps and farts and grindings, constantly reconfiguring the lie of the land, or the whole thundering cacophony of Mr MacDonald's Farm. Or, again, the ocean, dropping bombs on to the beach in the form of waves. When the surf is up, the Earth moves, quite literally.

The essential problem for LIGO or any other would-be detector of grav waves is this: everything is waves, vibrations in space, sinusoidal anfractuosities. Even particles, given half a chance, prefer to come not single soldiers but in battalions, foaming up into waves. So: how the hell do you tell the waves you want (especially infinitesimal ones) from the very big noisy bastards you don't want? The good vibrations from the bad ones? How, in short, do you get rid of the junk?

In a way the answer is simple. Keep going. It's just a question, as Jules Verne said, of whether health and strength permit you to finish the task. One by one, very gradually and progressively and patiently, you have to identify all the other waves, the misfits, and eliminate them one by one from your enquiries. This is what Mike Landry had been doing ever since he got here. When you switch on an optical telescope, it either works or it doesn't; with an interferometer, it still takes a lot of fine-tuning and doctoring and massage. George Smoot, looking for one part in 100,000 in the microwave background, said it was like 'listening for a whisper during a noisy beach party while radios blare, waves crash, people yell, dogs bark, and dune buggies roar'. Mike Landry would have given his eye-teeth for roaring dune buggies. Dune buggies were easy. Over the years, while trying to tune in to the pulse of the cosmos ('the stochastic wave background, our Holy Grail'), he had been troubled by: (a) the postman; (b) rabbits crunching carrots; (c) distant waterfalls. I wouldn't be surprised if a leaf falling or a snowflake smashing into the cold, hard ground could cause a major meltdown at LIGO. Mike had brought in instrument after instrument to track down and crack down on every stray extraneous disturbance: he'd started with a seismometer, a thermometer, and a magnetometer, and gradually had to get subtler and more cunning and come up with a rabbitometer and a leafometer to keep every little detail of the outside world under constant surveillance ('to know the external noise source').

I almost forgot to mention: the sun and the moon are another couple of potential ghosts in the machine, evil genies. You think they only pull the tides around? Think again. They distort the Earth's crust, which in turn truncates the building (or stretches it) by a couple of hundred microns (and one micron was already a million times more than the signal they were looking for). Sabotage. Worse than rabbits.

There were different angles of attack. 'Cleaning' the beam involved passing it through a silicon block, triangular in shape, that like a highly efficient border patrol stopped higher-order light smuggling itself in. A 'reference car' carried out a further customs check. Keeping it like an igloo in there froze out any hot frequencies. They had a handy little electro-optical tagging device that actually stuck a flag (known as a 'sideband') on the shoulder of the laser beams so they could ID the good guys. 'Control systems': LIGO was stacked with them.

One other answer Mike came up with went like this: 'If you're driving down the open road and there is a crosswind coming from the left, you're naturally going to turn the wheel to the left to try and compensate. You just have to know the speed of the wind and the force as it hits you.' So, in the same way, Mike fed negative images of waterfalls, rabbits, wild horses, the sun and the moon into the machine so as to balance things up and keep it going down the middle of the open road. LIGO was probably the most finely balanced detector on the planet, as poised and precarious as a tightrope artist edging over the top of Niagara Falls.

24

I didn't tell Mike Landry I had a secret reason for being there. I didn't tell anyone. I loved the physics, it was true, I wanted the truth. But, in a flaky off-and-on kind of way, it was also true that I had been writing a novel in which the hero had to escape from a lab something like LIGO. Like Forward, I was dabbling in fiction too. My lab had the same laser interferometer apparatus for detecting grav waves. Obscure forces (I was working on the detail) were determined to shut it down and prevent science from discovering the secrets of the universe. My hero (called Sven) found himself stuck in the basement, all on his own, with a couple of ruthless killers trying to get down in the elevator after him, and for all he knew a whole army surrounding the entire lab. The problem was: how did he get out? Or could he get out? Maybe he would have to die and the novel would need a new hero.

I handed the problem over to my two teenage sons, with a sketch of the scene, and asked them to come up with a solution. The best they could do was some kind of highly experimental invisibility device, a 'light-bender'. Ingenious, but at the same time too easy, too convenient. He just *happened* to have an invisibility belt to hand on the day a bunch of assassins, armed to the teeth, came to call? It was too Tom and Jerry, where you just reach beyond the frame to grab anything that's necessary. I needed something cruder and more effective. I had an idea, but I wasn't sure it was feasible. I had to get into the inner sanctum of LIGO to check. I had to break in to see if it was possible to break out again. Mike led me into the vast chamber, the size of an aircraft hangar, which housed the core of LIGO. The Tube Room.

I was wearing disposable shoes and a clean-suit. If I started sweating I was going to raise the temperature a few millikelvins and the whole thing would go up in smoke. Be cool. I hardly dared breathe for fear the additional water vapour would jiggle all the settings or blow the whole thing over. I really could have done with a face mask too.

What I was looking at was two identical 4-kilometre-long spaceships at 90 degrees to one another that have just docked. It was like seeing *2001: A Space Odyssey*, but for real. All that was missing was the 'Blue Danube' going wah-wah-wah-wah-waah in the background (but that would have thrown all the instruments off). During construction – by Chicago Bridge and Iron – a whole factory was set up on this site, first to build the tools to build the instruments, then manufacturing enough steel to house the lasers, millions of tons, miles of the stuff, which had to be shaped and welded and aligned and smoothed down and polished to perfection, like some kind of musical instrument, a very long flute or a didgeridoo. They started at the middle and grew it out at both ends, till they had eight solid kilometres of parabolic tube, like a giant toilet roll. The whole thing was then wrapped in foil and baked for 6 months, using a current to turn it into a huge microwave, to get all the water out and excite stray molecules. And then, finally, LIGO was let loose, set free from the Earth, like an angel, like a Saturn V rocket. The factory then took itself to pieces, and reassembled in Louisiana to repeat the whole process for

LIGO's twin sister near Baton Rouge, with the small additional problem of snakes and poisonous spiders to worry about.

There were two separate experiments going on inside the tubes, the first with two beams each 4 kilometres long flying away and then spinning back towards one another, like a pair of ice-skaters pulling off a difficult stunt; and another with beams 2 km long, half the length but otherwise identical. The way they reconnected with each other would spray up different interference patterns on the screen: every time the pattern changed it was tripping an alarm and there was the possibility that a wave that was first thrown up billions of years ago had just hit. But then they still had to check their results (a green line) with the twin down in Livingston (the red line), to confirm there was a match and it wasn't just a local anomaly, another rabbit munching a carrot or the postman dropping a letter through the door.

LIGO is an all-sky device, sensitive in all directions (although more sensitive in some than others). Mike thought of the two tubes as extremely long tuning forks, pitched to the frequency of the universe. 'The light is resonating in the arm,' he said. They not only stretched out the beam but sustained it as long as possible, storing it up in Fabry–Pérot cavities, like compost. It was what Mike called 'signal recycling'.

'Can I just stick my head in and have a look round?' I said.

'Ahem, there are laser beams bouncing around in there.' I think Mike was more worried about the beams than me: he was protective of his baby. 'And it's a vacuum.'

'It seems a pity, you know, having come all this way and all.' I was begging him with my eyes. Of course I couldn't let him know why I wanted to do it.

'Well, there is a power-down period. I guess you could look into the beam-splitter. It's disconnected from the vacuum section. You'll have to wear goggles though.'

I wore orange goggles. Great thick things with weird lenses that made me feel I was in the middle of some kind of interstellar struggle. Which, in a way, I was. Mike opened up, as delicately as a surgeon.

'The laser's infrared. But you can see the glow on the mirrors.'

I was looking into the heart of the beast. Through a glass, orangely. You could see the blood circulating, in the form of pure light. At one end: master oscillator; solid-state laser; amplifier; four photodiodes. The starting point was also the final resting place, where light returned to be digitized. If I turned my head and looked down the tube, miles long, I could see laser beams playing tunes from the beyond. It was like looking down the barrel of infinity.

'It's pretty roomy in there,' I said admiringly.

'Five-foot diameter.'

'So,' I said, trying not to give away too much of my plot, 'do you reckon you could actually hop in there and stroll about?'

'If you didn't mind the vacuum.'

'Good point,' I said. 'But assuming you had switched off the vacuum beforehand. Then you could get in? I mean, I can see you'd have to duck your head a bit.'

'Like the spring-clean? Yeah, we do it every so often. For maintenance. Not in the middle of a run though.'

'And, just say for argument's sake,' I went on, 'you got in there and you'd switched off the vacuum and you'd switched off all the laser beams bopping about, and you walked along the tube, could you get out again? Or are you stuck?'

Mike had to stop and think about it. He looked at me through his goggles. He must have thought I was mad. I probably looked mad with the goggles on. 'It's like a submarine. There are hatches along the way. They lead to an inspection door. So, yes, I suppose you could. I haven't done it, I don't think anyone has, but it could be done.'

It was enough. I could see how to get in, I could see how to get out. Sven, my hero, was saved. There was hope for me yet.

Michelson interferometer pattern

25

The first beam-splitter and interferometer was the brainchild of Albert Michelson and Edward Morley. They didn't have lasers in 1887, but they realized that light beams could constructively and destructively interfere with one another to produce a pattern. As far as they were concerned though, this wasn't the point: all they wanted to do was finally prove, once and for all, the existence of the ether. The ether was something like heaven: it ought to be there, somewhere, but no one was quite sure what it was or where it was or if it was.

Just as sound waves need air to propagate and waves need water, so it was assumed that light waves must be flashing through a medium – the 'luminiferous ether'. It was a reasonable supposition. As early as 1657 Otto von Guericke, the Burgomeister of Magdeburg, Germany, had carried out a graphic experiment. He put a ringing bell inside a glass jar and then proceeded to pump all the air out. The audience could still see the clapper hitting the bell but they could no longer hear it. Sound, therefore, could not travel through a vacuum. In contrast, light must, otherwise the bell itself would have vanished along with its sound and the bell jar would be dark inside. But all the same: something (light) travelling through nothing (the vacuum)? It was a paradox and science abhorred paradoxes. The luminiferous ether, a kind of diaphanous celestial mist, otherwise known as 'the First Postulate', was the answer. But if there really was such a thing as ether then it ought to impact on the speed of light. Michelson, obsessed by measurement, saw the Earth as something like a convertible car or a motorbike flying along the celestial road, with the wind blowing in your hair. If you were driving at 50 mph, then the air would be thrusting into you at the same speed. So too, in theory, with the 'ether wind', which would be proportionate to the speed of the Earth (some 100,000 km/h).

At the age of 25, after studies in Berlin and Paris, dissatisfied by the fuzziness of existing approximations, Michelson had set about definitively measuring the speed of light, in 1878 at the US Naval Academy in Annapolis (where he had once been a midshipman, but better at optics than anything else). Using a variation on Foucault's method, he bounced a beam from an arc light back and forth off rapidly rotating mirrors (250 revs per second), driven by a large turbine, along a path 2,000 feet long by the side of a sea wall. The reflection off the spinning mirror, on its return from the journey down the tunnel, would hit a slightly different spot. After a year or two of experimentation, the difference enabled him to calculate the speed at 299,940 km/s, plus or minus 50, a figure 20 times more accurate than all previous estimations. He was buoyed by this early success. Surely it couldn't be that hard to crack the ether problem too, and pinpoint the speed of light when you added on or subtracted the relative motion of the Earth? Michelson brought in Morley – a chemist but also a practical experimenter of renown – because he was struggling.

Admittedly, it didn't look much like a convertible. Michelson and Morley built an octagonal brick pier in a basement room at the Case Institute of Technology in Cleveland, Ohio, and covered it with cement. On that very solid base they placed a circular iron trough and filled it with mercury. They placed a wooden float, a foot thick, in the trough. Then they put everything else on top of the float: an Argand burner (to act as light source), a half-silvered mirror to split the beams, and then lots of other mirrors to bounce the beams to and fro until they finally pitched up on an interference screen. Everything had been scrupulously honed, polished, oiled, measured, levelled, aligned, then taken apart and put back together again and reconfigured, just to make sure. They flipped a switch on their light source and sent off twin light beams at right angles to one another, flashing around their circuit of mirrors. They weren't quite sure which direction the ether wind was blowing in, so they slowly revolved the equipment in the bath, one turn every six minutes. In any one rotation each arm would be facing into or away from the wind twice. One beam would be going with, and the other against the flow, swimming upstream or downstream, and therefore travelling faster or slower accordingly, very slightly out of phase, producing a measurable difference in the interference pattern. It was a brilliant and fanatical experiment.

There was only one small problem: all the beams promptly returned to their starting point, like dogs returning from their daily constitutional, at the exact same time, regardless of whether it was day or night and even what season it was. No significant difference at all in V for velocity (nothing at least that didn't come under the heading of experimental error). No change of pattern. No visible peaks and troughs. No wind rushing through the hair. No ether-induced lag. Ergo, no ether, either. It looked strangely as if something really could pass through nothing. The experiment was a glorious failure. Michelson and Morley couldn't quite believe it, though, and went on coming up with ever more sophisticated pieces of equipment, more mirrors, longer paths, heavier blocks of stone, but the result, however they sliced it up, always came out the same.

Their 1887 paper, published in the *American Journal of Science*, still clung valiantly – after seven years of negative results – to the validity of the First Postulate. The two men theorized that the laboratory walls must be blocking out the ether or that the Earth's gravitational field dragged the ether around with it, cancelling it out ('entrainment'). They conjectured that they would get better results if they could take the whole apparatus and stick it 'at the top of an isolated mountain peak'. Harking back to Rømer, they suggested bringing into play the eclipses of the moons of Jupiter. Perhaps it would work better in a vacuum? They refused to give up or lose hope: 'the difficulties are entirely mechanical and may possibly be surmounted in the course of time'. It was a classic never-say-die attitude. If health and strength permit me. Michelson would sometimes go days without eating or sleeping. When he slept he had a recurring nightmare about riding a motorbike up an endless hill, his hair flying in the ether wind. He measured the diameter of a star. He fired light rays off the top of Mount Wilson. His original experiment, with infinite refinements and variations, lived on.

It was Einstein who assumed the role of grand executioner. The young Einstein wasn't interested in the ether – from his point of view, the First Postulate was just another of the grand illusions of nineteenth-century science. It was 'superfluous'. He was benevolent enough to allow that the existence of the ether could never be fully ruled out, so long as it was stripped of any measurable real qualities and effects. You can have it, he seemed to be saying, so long as it has no substance whatsoever. Heaven, but with nothing and no one in it. But he retained from the apparently failed experiment the idea that light travelled at a constant speed, come what may, ether or no ether, with or without the velocity of Earth, and all his thinking in special relativity was guided and inspired by that one point.

Michelson accepted Einstein's work, but all the same he never gave up his belief (according to his daughter and biographer) in the ether. He was still testing the speed of light, and perhaps hoping to find faint traces of the ether, in 1931, at the Irvine Ranch in Orange County, this time in a tunnel a mile long with a mirror spinning 500 times a second, when he was struck down and killed by a cerebral haemorrhage. Thus his most successful and well-remembered experiment was a failure. Sometimes a negative can be a positive.

Even with their 4-kilometre-long beams – a direct descendant of the Michelson–Morley interferometer – Mike Landry and Fred Raab and Evan were still looking for the tell-tale shift in the interference pattern as the two beams collided. Waiting for something to *interfere* with the length of the laser beam. Something like gravitational waves rising and falling as they hit the Earth, peaks and troughs, subtly stretching or squeezing the two beams. So far one would have to add 'imperceptibly'. But the whole point of LIGO was to perceive the imperceptible. Throw a lasso around the horns of those invisible cows.

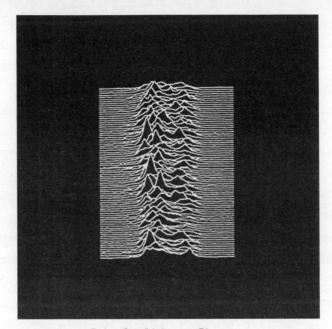

Dying Star/Unknown Pleasures

26

I took off the protective clothing and dumped the disposable shoes.

'You wonder, when you come to work in the morning, what the science team has seen overnight.' Mike sounded calm, hopeful, anything but despondent. 'But it wouldn't be astounding. We're not going to go running around in hysteria. If we were to find something. We stay cautious. We'd have to sit around and argue about it. It's likely that the signal will be weak to begin with. At the limit of sensitivity.'

Probably most people would have given up by now. If they didn't have any observations soon, it was going to get harder to attract enthusiastic young grad students like Evan. There was always going to be a lot of competition from more established fields like particles. But Mike was an optimist. They were all optimists. Keepers of the faith. They were like a bunch of surfers paddling around in a flat sea, convinced that there was a big swell on the way (except in their case it was the exact opposite of a big swell – the movie would have to be called *Small Wednesday*).

Evan had said, 'If we could build one a hundred kilometres long, we would. For a given wave, the further apart the two end points the greater the change in distance.' Mike didn't think it was that easy. 'It's counter-productive after a certain point,' he said. 'You've got the curve of the Earth to worry about. The longer the arm, the more you have to hang the test masses at an angle. Then you have to compensate. And that makes noise.' He wrinkled his nose as if something unpleasant had just wafted by. 'Four to six kilometres is the max. On Earth.' In the realm of grav wave detection, size was, apparently, not quite everything. 'That's why we have LISA.'

Correction: size *was* everything after all. LISA (Laser Interferometer Space Antenna) was next-generation. It got around voluptuous terrestrial curves and all that human-engineered racket by orbiting around the sun, but trailing the Earth by around 20 degrees. And the vacuum was free up there. You didn't need immense toilet rolls to wrap around the beams. LISA didn't exist yet, except on the drawing board. It had arms not 100 km long, but 5 *million* km. It would be the longest straight line ever created. And there would be three of them, defining the greatest equilateral triangle ever seen in the history of the universe. Nicholas of Cusa, the Renaissance mathematician and philosopher and inventor of concave-lens spectacles, wrote that 'An infinite straight line would contain a triangle, a circle, and all possible geometries.' LISA was as close as we were going to get to that infinite straight line. The triangle was the star-gate through which information about all possible geometries would pour, the sum of all histories. It was the eye of God.

The Earth is so noisy that if you want true peace and quiet you have to hop on a rocket. But I realized then that if LISA was LIGO spun up into space, then LIGO was, conversely, a terrestrial LISA, doing its level best to float free of its earthly constraints. 'The slab we're standing on,' said Mike proudly, 'is decoupled from the rest of the lab.' It was, if not quite orbiting around the sun on a separate trajectory, at least swaddled up in a seismic isolation system, cushioned several times over with layer upon layer of polymer damping, suspended on springs, surrounded by a cryogenic moat fed by liquid nitrogen tanks, deeply insulated and sheltered from the turmoil and turbulence of raw existence and basking in the most profound vacuum we have yet been able to concoct. 'We're trying to make it as if we were already in space. In free-fall.' The tubes really were like starships, it wasn't just my sci-fi fantasy.

Mike took me through another door and suddenly we were outside, back in the light and the air, but looking out over a landscape that seemed to belong to another planet. Arid, barren, and illuminated by the glare of exotic stars. I slipped my shades on. The north-western tube stretched out ahead of us like an elongated molehill, all 4 kilometres of it, across parched sands, reaching to the foot of distant purple hills. I could see the inspection doors at regular intervals along the length of the tube. I had envisaged Sven emerging from one of them and taking flight, on the run from his would-be killers. The question was – where would he go *to*? The terrain was flat, open, and inhospitable. He would be a sitting duck for any hostiles. I had to get him not just out of LIGO but out of Washington state. My two sons – I'd asked their advice again – were suggesting a passing flying saucer.

Just as I had a secret, so too did Mike. But he confessed. 'You know,' he said, before we parted, 'we're not really astronomers.'

'No?' I was a little shocked. 'What are you then?'

'Well, we're on the fringe. We're mostly engineers. Or particle physicists. Or general relativists. We don't study neutron stars for a living. If astronomy was Hollywood, we'd be the struggling indie part of the industry.' LIGO was New Wave. It was guerrilla filmmaking. They didn't have any well-known stars to speak of. Or a bottomless well of money. It was Tarantino or Robert

Rodriguez in their most radical, risky, marginal phase. They were mariachis, in their different ways, with a common fondness for highly violent incidents, whether on Earth or elsewhere. And I realized this: that violence can be creative as well as destructive. And that the greatest act of creation was also the most violent.

If I had been asked what Mike's favourite recording was, I would have guessed Simon and Garfunkel's 'The Sounds of Silence'. The one he actually mentioned was Joy Division, *Unknown Pleasures* (1979). The thing about it was the cover: a picture of the first pulsar, CP1919, discovered by Jocelyn Bell back in 1967 in Cambridge. CP1919 was a neutron star: a large star that had started to run out of fuel, over-inflated, popped (the phase known as supernova), then shrunk back down to an insanely dense diameter of 20 km. Now it was spinning very fast (once every 1.337 seconds) and zapping out a signal every few seconds as it swung by, like a very distant lighthouse. The picture of these consecutive pulses is referred to as the 'dying-star logo'.

'Gravitational waves could make a fantastic record cover,' said Mike.

27

After I got back to Cambridge from LIGO, I discovered that my old friend Sidney was not the only one to have died recently. Far harsher than any thriller you care to name, the reality was that thousands were dropping dead every day, and they were all assuming the form of light, and could be viewed at any time on various distant planets. But the only other one I knew personally was Richard Rorty. He was a professor of comparative literature at Stanford University. But he was a philosopher by training and inclination. The BBC phoned me up and asked me to speak about him in a programme called *Last Word*, a kind of radio obituaries column.

I had been thinking of Rorty because he didn't really believe in mirrors. Or rather he believed in their existence but he didn't think we should make quite such a big deal out of them. In his influential book *Philosophy and the Mirror of Nature* (1980) he mocked the attempts over the ages of philosophers and writers and people at large to represent what they were doing as mirroring nature, as Shakespeare put it, 'holding the mirror up to nature'.

Stendhal, the French novelist, said that the novel was 'a mirror taken for a walk along the street' (*Le Rouge et le noir*). Wittgenstein, in the *Tractatus Logico-Philosophicus*, argued that language ought to mirror reality in its structure, and spoke rather reverently of 'the great mirror' (5.511). But according to Rorty, poking fun at any idea of our 'glassy essence', this is all just another fanciful and misleading metaphor, perhaps the mother of all metaphors. For Rorty, as for Nietzsche, language was nothing but a bundle of metaphors, metonyms, and anthropomorphisms. So the best you could do was be ironic about it (the ironist taking over from the metaphysician). There was no truth, and realism was just a fallacy. Rorty – together with the whole of postmodernism – was anti-mirror, pro-speculation but anti-specular (somewhere, some time, he still is). The mirror, as Borges would say, was abominable because it multiplied the number of men. Rorty would agree with him that 'mirrors have something monstrous about them'. For him, every time we talk about mirrors we are just talking about ourselves, our obsessions and delusions. In this sense, everyone is a little like Narcissus.

I met Rorty only once (although I went along to a few of his lectures over the ages and read several of his books). He was from New York but he put me in mind of Garrison Keillor, author of *Lake Wobegon Days*, the great Midwest pastoral and lament. He had a definite Lake Wobegon look about him. He was laid-back, droll but deadpan, with a melancholy air that hung about him like the little cloud around a smoker. We were sitting in some senior common room, surrounded by old dons drinking coffee (in some cases something stronger) and reading the newspaper. I said to him: 'So if someone put a gun to your head and cocked the trigger, wouldn't you be more or less concerned according to whether she said "Don't worry, this gun is unloaded," or "It has six bullets in and I'm going to use all six," or "One bullet only – spin the barrel and let's see what happens"?'

It was the least likely thing in the world to happen at that precise moment. I was asking him to imagine this scenario because he professed an extreme scepticism about all language. But I thought that if he was more or less concerned by one or other of those statements, then he must be taking them fairly seriously. He thought about if for a while and then replied: 'It seems to me like the gun is the main thing in this: I would be concerned about that, I admit. But everything else is just shooting the breeze.'

He was a funny guy even if he never laughed. And he had a good point because if someone was saying those things to you, you wouldn't necessarily know what to believe. On the other hand, it would be reasonable to take precautions (duck, run, karate kick, pull out a gun of your own, etc.). But, now I look back at it, the key thing was *seeing* the gun. We talk of the mirror as being like an eye, but the eye is a mirror with a brain behind it to make sense of the image. The telescope struggled with the atmosphere and astronomers were trying to take the twinkle out of the star. Rorty was saying: the atmosphere is really all there is. We might as well enjoy the twinkle. He had no notion of trying to 'resolve' anything. He didn't take scientists too seriously. He thought of science as all tell and no show, like Marcel Proust with very expensive toys.

I used to sympathize with Richard Rorty's argument. But now, after visiting the Keck and LIGO, I couldn't help but feel that he just hadn't looked into mirrors closely enough, he just assumed that a mirror had to be a metaphor from the very beginning, supposing you could work out what the beginning was in the first place. I thought it was reasonable, given how much mirrors had already achieved, to try to mirror mirrors. Mirroring is just what we do. Mirroring is what Rorty used to do all along, he just didn't like what he was doing. Actual mirrors are an extension of ourselves. We are our mirrors. Rorty wanted to smash a few mirrors, just to defy the superstition.

Maybe Fred Raab felt the same way about mirrors. Mirrors were old hat, mirrors were for moths. We needed to move on from mirrors, go *through* the looking-glass.

Rorty and I, having that brief conversation about the chances of being gunned down, can now be seen through a very powerful telescope on a planet some twenty light years from here. (Like Hal in *2001: A Space Odyssey*, you would have to be able to lipread to make sense of what we were saying though.)

28

Albert Einstein and I had a couple of things in common. For one, he didn't like socks (I discovered this at the Jewish Museum in Berlin, which had a photograph of him at a conference in a suit, good shoes, but no socks). For another, he had a sneaking regard for mirrors. He often looked into them and found them quite useful. Slovenly though he was, he still needed to shave and trim his moustache. In 1896, at the age of sixteen, before he even had a moustache, he carried out a thought experiment: what would he look like, he wondered, if he held up a mirror to his face while travelling at 300,000 km/s? He imagined that he was riding on the back of a beam of light. Would he even be able to see his own face? Light from his face would try to hit the mirror and produce a reflection, but it would never get there or even get away from his face, because it was already travelling at the speed of light. The mirror, at light speed, became unattainable. He would know when he was moving at the speed of light because his reflection would fade and vanish from the glass like a ghost. Light would become a frozen wave.

It was a beautiful and tragic image. The only problem was, Einstein eventually decided, that none of it made any sense. He couldn't believe that his reflection would disappear, that a mirror just wouldn't work any more, no matter how fast he was going. He realized that he had been seeing himself as travelling at some fixed speed relative to the ether. He concluded that the ether could not exist. Either the mirror or the ether had to go and he stuck with the mirror. The conclusion was obvious. If he could still see his reflection, then light must continue to move at 300,000 km/s *relative to the observer*. There was nothing else for it to be relative *to*. The observer becomes a very significant part of the equation. Just as there is no reflection in a mirror unless I hold it up to my face and look in it, so similarly light only really acquires a speed when it has someone around to see it and measure it. Galileo, Rømer, Michelson and Morley: they were doing more systematically what the rest of us do without knowing it or thinking about it. But no matter how fast we are going, or how slow, the result always comes out the same for each one of us. Light is strictly egalitarian, it won't move any faster just for you, no matter how much money or power you have. It belongs to each and every one of us. As do our own reflections. It was something else we had in common, he and I. We can see why it was the FBI feared that Einstein had radical socialist sympathies and kept tabs on him while he lived in the United States.

But if light is a constant then Michelson's convertible analogy won't work any more. If you throw a cigarette butt out of your convertible and hit a woman in the face driving her convertible in the opposite direction and you are both going at 50 mph, then the cigarette hits the woman at 100 mph. The speed of the butt is the sum of the two speeds. But if you are just looking across at the woman, there is no difference: light travels from her to you and from you to her at the exact same speed, with no added extras. This is why Michelson and Morley faced disappointment: they were hoping to add or subtract some speed to or from light and measure the difference. But there is no difference. Thou shalt not add nor subtract where light is concerned. In this sense, light is God in Einstein's theory. Light had to be a constant, otherwise the reflection would be lost from the mirror. Which could not be allowed. Whether he was standing still in front of his bathroom mirror, or flying through the air at light speed holding a mirror up to his face, the reflection would remain the same, in both cases slightly younger than Einstein himself by a matter of a few nanoseconds.

In 1905, aged twenty-six, now a young father, working in the cloistered calm of his patent office in Berne, Switzerland, Einstein published three short but beautiful papers in the *Annalen der Physik*. The first analysed Brownian motion and showed that matter must break down into molecules, not just atoms, and that you could actually *see* them: particles 'perform motions of such magnitudes that they can be easily observed with a microscope'. The second, 'On a Heuristic Viewpoint Concerning the Production and Transformation of Light', considered the quantum nature of the photoelectric effect and introduced the concept of (if not yet the word) the photon. The third, 'On the Electrodynamics of Moving Bodies', which opens up the sphere of relativity, developed his thoughts on light. The paper is full of mirrors, reflecting surfaces and rays bouncing back to their source. No wonder that Einstein was concerned, too, with the pressure of light on mirrors. Like Michelson, he could hardly believe his own theories. 'I must confess that at the very beginning when the special theory of relativity began to germinate in me,' he wrote later, 'I was visited by all sorts of nervous conflicts. When young I used to go away for weeks in a state of confusion, as one who at the time had yet to overcome the state of stupefaction in his first encounter with such questions.'

If it is our relationship to light that is fundamental, then our relationship to time and space becomes secondary and much more fluid than previously supposed. For Newton, the whole universe ticked along in time with a single clock. Einstein has all his observers carrying their own clocks around with them and finding that they no longer coincide with other people's clocks. Everyone is a little out of sync with everyone else. If I've understood Einstein correctly, Jung's theory of synchronicity is a mistake. Maybe that explained why he didn't really believe in entangled particles: simultaneous transmission of messages must be wrong for the simple reason that there is no simultaneity. Riding a streetcar through Bern, Einstein looked back at the clock tower that dominated the city. He realized that if the streetcar were to take off at (or near to) the speed of light, the hands of the Bern clock must appear to Einstein, inside the streetcar, to stand still, whereas the second hand of the watch on his wrist would still be ticking around as usual. My time is distinct from your time.

It was like Galileo's old longitude problem revisited and expanded. We could define a 'prime meridian', 0 degrees, and locate it at Greenwich (or wherever), but as we travel around the universe we will no longer coincide with it. Just as there is no absolutely stationary space, nor is there any absolute cosmic clock any more, not in the moons of Jupiter or anywhere else. Even on Earth 'a balance-clock at the equator must go more slowly, by a very small amount, than a precisely similar clock situated at one of the poles under otherwise identical conditions'. The original point of reference is purely arbitrary and all travellers – and spacemen especially – have their own clocks, all set to different time zones. The faster I go, Einstein concluded, relative to a particular observer, and the closer I approach the speed of light, the slower my clock appears to go, from the point of view of that observer. If I return to Earth after a very quick tour around the galaxy, I will discover all my friends are dead and gone while I am still in the prime of life. If the extremely ugly man in the *Astounding Tales* story takes a return ride on the magic bus (or streetcar), all those who reviled and abused him for his ugliness will no longer exist.

The Hulse–Taylor inspiral was like a relativity experiment in a distant laboratory. The two dense miniature stars swirling around one another emitted a chronometric pulse, one every 0.059 seconds. Even pulsars carry their own clock around with them. But there were points in their orbits where the gravitational field was strongest and they speeded up, and points where it was weak and they slowed down. It was found that, just as Einstein predicted, the ticking slowed down, from Earth's point of view, when they were going fastest – the 'relativistic time delay' – and sped up again when they slowed down. QED.

If I could only attain the form of light, travelling at the speed limit, then the clock would stop completely and I would live for ever. Thus immortality is attainable, it seems, but only by light itself. Perhaps because Einstein understood that humans could not become immortal while they still had mass, he argued that we could never reach the speed of light. Only our images are made in the image of God. I read all the papers and my rusty A-level maths had to be hugely supplemented by Unc's hints and tips. But even I could see that light penetrated into all Einstein's equations. Getting close to light tweaked things. A rod decreased in length as it approached the speed of light (shorter by $\frac{1}{\sqrt{(1-\frac{v^2}{c^2})}}$); time slowed down (it was slower by $\frac{1}{\sqrt{(1-\frac{v^2}{c^2})}}$)

Time and space – as Minkowski recognized, following Einstein – were effectively interchangeable, symmetrical, different states of the same object, since they both hinged on light. If all information depended on its transmission by light, then matter and energy too were already inhabited by light. Almost as a footnote, Einstein published yet one more paper in September 1905, 'Does the Inertia of a Body Depend Upon Its Energy-Content?', which spelt out for the first time the full implication of the earlier paper. The central equation ($m = L/c^2$) would eventually take the form $E = mc^2$. Energy equals mass multiplied by the square of the speed of light. The symbols of ultimate power. Power squared.

Where Galileo maintained that the universe was like a book written in the language of mathematics, Einstein had a more post-Babel-style conception: 'a huge library, whose walls are covered to the ceiling with books in many different languages'. The books were similar to one another, but not exactly the same. They were all part of the same library, and yet they each told a different story in different vocabulary, at different times. This was the quantum side of Einstein – he didn't expect to have to understand everything. Nor could anyone understand fully what was going on in someone else's frame of reference (except you knew that they could always see their reflection in a mirror). But there was a classical side to Einstein too. He wanted to be another Newton and pull together all those frames of reference, translate all the books into one unified language. Having injected weird subjectivity into time, he wanted to snatch it back again and smooth things out.

The General Theory of Relativity, of 1915, incorporated all possible moving bodies, whether they were moving at constant speed (special theory) or accelerating or decelerating. It rethought gravity – defined previously as an obscure force – as a form of curvature in space-time. Gravity had a shape. Very massive objects actually warp the geometry of the universe, they curve space-time, like distorting fairground mirrors. And very massive astrophysical events, therefore, produce tremors in the curvature, with amplitude – gravity waves, which travel at c. Gravity could bend light, it could even black it out entirely. Maybe, therefore, it could come up with answers when light flickered and faded and died.

29

One night, somewhere in the wilds of Australia, Spencer (aged four and a half) and I were sitting out on some cliffs staring up at the night sky. We were about a thousand miles from the nearest city. The Milky Way was right in our faces. It was snowing stars. A blizzard of light. When we finally decamped and went back to a motel he was inconsolable and cried his eyes out, as if I had ripped him away from his own mother.

I knew just how he felt, the time I looked up and saw the flying saucer.

It was late at night. But it was a clear night and my head was clear too. I had had only one glass of orange juice to drink all evening. So we can rule out alcohol-induced hallucination. Nor was it a Hollywood-inspired fantasy: I had *not* been to see *E.T.*, *Alien(s)*, *Independence Day*, *Close Encounters of the Third Kind*, *Invasion of the Body Snatchers*, *2001: A Space Odyssey*, *War of the Worlds*, *Mars Attacks!*, or any one of a thousand low-budget movies called something like *The Thing from Planet X*. Nor had I been bingeing on reruns of *Star Trek*. As a matter of fact, I was returning from a school concert. It had finished with a rousing rendition of Eric Coates's 'Dambusters March', which might just have made me more likely to survey the skies. And, admittedly, there was a full moon that night (a Friday night, to be precise). But rather than induce a moment of insanity – lunacy – it only made what I saw all the more clearly visible. I had no particular reason to see a flying saucer. I was not even – for once – scanning the skies, eagerly, wondering what was out there. My thoughts were exclusively terrestrial. I was thinking of nothing more exotic or astounding than getting home, making a cup of tea, maybe eating an apple, and going to bed. I was busy, I had a lot on, there was work to be done. I was in no mood for alien abduction.

As I cycled up Newton Road (in Cambridge), my first thought was: what idiot has just put up a mobile telephone mast several hundred feet high and stuck a bunch of fairy lights on top? Whatever happened to planning permission? For a couple of hundred yards, I had a clear view of a trio of radiant white lights in the sky, garnished with a few blinking blue and red lights. What I couldn't work out was what they were doing there. I was pedalling while they remained perfectly still. They weren't going anywhere. They were as stationary as a lighthouse perched on an invisible promontory. Obviously, they were nothing to do with an aeroplane. It was odd, that much was clear.

As I approached the end of Newton Road and my house, I got off my bike and dropped it right there on the ground. I tipped my head right back and – slack-jawed – turned my gaze almost straight up (up or down or out: let us say, vertical). Something to do with shepherds being guided by a star flashed through my head. But in fact it was more starship than star, the bright lights just hanging there in the sky, I would estimate at a height of around four or five hundred feet. And here is the strangest thing (one of many strange things): there was no noise, not a boom, not a zoom, not a whisper. As I looked more closely at this celestial object, I could see the dark outline of a craft that was nothing like a plane. No central tube, no wings jutting out at the sides; more . . . rounded. But it was big, definitely Jumbo-sized. And it wasn't going up and it wasn't coming down. It was going neither right nor left. It wasn't doing anything, it just *was*. I am only describing what I saw, no more and no less.

And then the thought occurred to me: I haven't seen anything as uncanny and incomprehensible as this since the day that France beat Brazil in the 1998 World Cup final. And oddly enough there was a faint resemblance between the thing floating in the ether somewhere over my head and the Stade de France in Paris. But this was a football stadium several hundred feet up, elegantly poised in mid-air.

A couple of minutes passed by in which nothing – except for this great miraculous nothing in the sky – happened. Then, smoothly, gently, quietly, unfathomably, the UFO – for it was certainly unidentified and it *was* a flying object, so what else are you going to call it? – seemingly tired of immobility and pulled away from its parking spot, like a bus leaving the bus stop, picking up speed as it went over my head and over the house and across the back garden in the rough direction of Girton, miles away to the north. An echo reached my ears of a faint rumble – a kind of deep purring noise – as it went by. The way it moved had a silky, dancing, unpredictable quality, like a Brazilian winger (but of a kind that could not be blocked by any French defender). I ran a little way down the next street to get a better view, but it had already vanished. Probably for ever. I felt like shouting 'Come back!' In fact, I think I did shout 'Come back!' I may have clasped my hands together, as if in prayer. I felt a terrible, crushing sense of loss.

If anyone wanted further confirmation, I could offer my wife and Spencer (who had been playing in the concert) as witnesses. They were with me at the time. While I had been staring at the sky, I had forgotten all about them. I had forgotten they existed. For a brief and immemorial moment I had forgotten I existed. Now, as I trudged back up the street, feet made of lead, I saw them and recalled who they were and said to them, 'Was that a UFO?'

'Yes,' said my son. 'Is there any chocolate going? I'm starved.'

'Yes,' said my wife. 'It's past your bedtime now.' (I don't think she was talking to me.)

They wisely got on with their lives. They went about their business. They never forgot that they existed. But for at least 24 hours I was mentally paralysed by the knowledge that there is, indeed, extraterrestrial life, and some of it had been visiting the city of Cambridge. There was nothing the least bit terrestrial about that flying vehicle, I was convinced. I hadn't been abducted and taken off to a better place or subjected to overly intrusive physiological experimentation, it was true. (If I had been, you might find this event even more difficult to take seriously.) But without doubt, this was a full-on encounter with an emissary of an advanced civilization. It may not have been a saucer, but it was definitely flying or, what was far stranger and even more emphatically alien, not flying but just sitting comfortably in the heavens, going nowhere, like an old man in an armchair with pipe and slippers.

Do I believe in aliens? Their very existence remains speculative. The hard evidence is minimal. But the statistics are overwhelming. The Hubble came up with an estimate of 125 billion galaxies out there a few years ago. A more recent estimate upped the ante to 500 billion. Assuming an average of around 100 billion stars per galaxy (I am using our own as a model), that adds up to a lot of suns. You only have to look at the galaxy-sized gas clouds out beyond the Milky Way and you can actually *see* the proteins that living beings could be *and were* made out of. There is masses of the stuff just floating around out there, like spare parts: it would be strange if it weren't scavenged up and reassembled into something – or someone. It is hard to believe that humankind is alone in contemplating the way things are stuck together.

Admittedly, though, my previous experience of alien life forms was fairly limited.

Unc, of course, had tried telephoning flying saucers, but nobody returned his call. Long before my own close encounter, I had happened to attend the annual conference of the British UFO Society, taking place at the University of Sheffield, on behalf of *The Times* newspaper. The star turn was a 'secret' documentary film of a post-mortem being carried out on a couple of dead creatures from outer space after their vehicle had crash-landed at Roswell in New Mexico in 1947. Roswell was legendary among ufologists, the subject of countless cover-ups. But – alas – it seemed that was all it was: a legend. Even committed card-carrying believers had to agree that the deceased in the film bore a strange resemblance to inflatable rubber dolls. You could almost see the air coming out as the scalpel sank in. The special effects were cheap and utterly unconvincing. Post-conference, however, at the pub across the street, the more people had to drink and the more they yarned on, the more extraterrestrial contacts they turned out to have had, until by the end of the night I was about the only man in the bar who had not personally had sex with an alien. Despite this, I remained sceptical. It struck me, when I did read about apparent sightings, that they tended to occur in obscure, isolated places, off the beaten track, somewhere on a deserted road in the wilds of East Anglia (or similar twilight zones in America, Mexico, or Russia), whereas one good landing in the middle of Trafalgar Square would have settled the argument fair and square.

But that Friday-night revelation changed everything. I was finally a believer. Aliens were simply real, not made out of celluloid and paranoia. Saturday night found me at a college party telling anyone who would listen about my otherworldly experience. I didn't care how crazy it sounded, it had happened, and there was no way around that brute, inexpungable fact. The next day I phoned up Unc (using an ordinary telephone) to tell him the good news. I was triumphant.

'What do you reckon?' I said. I should have known it was a mistake to ask this question.

'Where did you say this was?'

'Newton Road, right outside the house!'

'How far is Bassingbourn from you?' Unc said.

I didn't like the sound of it. What did Bassingbourn have to do with anything? 'I don't know, ten miles or so.'

'You know there is an experimental-aircraft facility there, don't you?'

'No, I don't know that.'

'Well, there is. Run by the Americans. I've heard they're experimenting precisely with noise reduction.'

'Bully for them,' I said, furious with the Americans.

'And they would tend to fly at night, I imagine,' Unc went on, 'to keep things hush-hush.'

I didn't want to, but I got his drift. Wasn't it more likely that the mysterious vehicle came from Bassingbourn than from Andromeda?

'Hey,' I said, remembering a crucial point. 'What about the *stationary* factor? Have the Americans come up with a way of anchoring an extremely large and massive object in the sky for an extended period? Have they?'

'Hmm,' Unc said. He had to admit that he hadn't heard anything about that, it was true. And he didn't know of any aircraft, experimental, American, or whatever, that could pull it off. But, on the other hand, that didn't mean it wasn't already happening.

I was cast down. I ceased to babble furiously about 'entities' and life 'out there'. I was not quite a broken man, but my newly found cast-iron certainty was shaken. There was a risk that as I looked up at the UFO in awe, a bunch of American airmen were looking down and laughing their heads off. Unc had crushed me yet again.

Looking back on the incident now, in the cool, uneventful light of day, my gaze fixed only on the 12-inch screen in front of me (just occasionally derailed by the swaying of trees through the window), I am tempted to think that Unc was probably right and the UFO was terrestrial rather than extraterrestrial in origin. But I don't know for sure. If this was a Hollywood movie, Unc would almost certainly turn out to be one of *them*. And maybe those mysterious phone calls were right and he wasn't my twin brother after all.

30

I was due to see the Godfather. Surely if anyone had the answers he would. But the Godfather was late.

An old hand at rocket science, cosmic strings, and wormholes, Kip Thorne had long been the chief flame-carrier and flag-waver for gravitational waves. He had actually studied under Joe Weber, in the midst of his wave rush, at a summer school in some mountain fastness in the French Alps. He hoped to see LISA elevated into orbit around the sun. He knew everyone. He knew everything. He was an international mover and shaker, a fixer and troubleshooter, who mingled with the Russians and the Chinese. And he had written at least one great book (*Black Holes and Time Warps*). I was scheduled to see him at Caltech (the California Institute of Technology) at 9.30. His secretary had arranged a half-hour slot for me, in between seeing senators and flying to Washington for lunch with the President and filming with Martin Scorsese (or something like that). I felt duly privileged that he had managed to squeeze me in. I had put on a tie that morning in his honour. I was hanging around in the cloistered corridors of Bridge

II, the physics building, on the Caltech campus in Pasadena, Los Angeles, where all the wave data were being sieved and sifted. The HQ. I felt as if I was finally homing in on the truth about everything.

But Kip Thorne had a toothache. Or a car accident. Or his wife's cousin had just died. Or all of the above. I forget what it was exactly. He did tell me, but I was in too much of a daze to take it in. I was slightly star-struck. Maybe it was something to do with the word 'Godfather' that people had been using about him. Not that he reminded me of Marlon Brando (he looked more like a reformed hippy), rather that the word broke down into 'God' and 'father', and he was old enough to be my father, and I probably slightly mixed him up with God too. (Or, as he put it, in slightly less theological terms, getting a grip on grav waves would be like 'being a creature from a higher dimension looking in on our universe'.) So our conversation in his office was a little limited. Particularly since the half-hour got shrunk down to ten minutes. That's how it goes when you are talking to God or a creature from a higher dimension. You walk out on the street and you remember all the things you should have asked.

I was staying at the hotel on campus, called the Athenaeum. I knew a place called the Athenaeum was going to be too expensive and it was. I had to get out of there and they booked me into another place that turned out to be even more expensive. On that particular day, I was worrying about flights and money and whether I had enough time in Pasadena to see everyone I needed to see. In fact, I was worrying about quite a lot of things, as follows, of no transcendent importance and in no particular order:

What! you have to pay for breakfast too? (I knew I should have booked a room at the EconoInn.)

How can I charge up my electric razor, having forgotten to bring the connector with me?

Is Spencer revising enough for his exams?

Is this damn mole on my lip some kind of skin cancer or what?

Am I gradually going deaf, like my father? Maybe I should go and see a doctor?

Looks like being a hot day, do I really need a jacket?

Viruses – is one coming my way today? Or is it already burrowing into my immune system? (I'd had two doses of flu a couple of months before and I was superstitiously expecting the third to hit any time.)

Where is the beach from here anyway?

Presently, within just a few hours, I was going to be regretting diving into a swimming pool with my mobile phone in my pocket, which – as you might expect – pretty much killed it. But for a long while I would be left wondering whether it could be resurrected with the correct loving care and attention. (Answer: no, it couldn't.)

And so on. A wave of worries or waves of worry. And that was without taking into account the usual background anxieties to do with dying, not dying, getting paid, getting taxed, and the possible (and in the long term inescapable) annihilation of the entire planet. I was worrying about everything.

I generally try to bracket out these considerations when I am thinking about the origin of the universe. In fact I probably try to think about the origin of the universe *in order* to bracket out these considerations. But they were brought forcibly back to mind by Alan Weinstein when he said, 'I don't understand my own wife. I have no idea why she does what she does.' He was talking about complexity. We all suffer from complexity. Alan Weinstein, Kip Thorne, and I – and every other earthling – lead complex lives. We have to negotiate our way around a hundred problems a day, most of them insoluble, just to survive. Above all we have to deal with our fellow human beings. And if we are honest with ourselves, we have to admit that other people are a bit of a mystery, even if we do have a lot of DNA in common. It's tough enough understanding why I do what I do. But as to what you or she or he or they are doing – who knows?

Alan Weinstein was in charge of baby LIGO at Caltech. I went over to see him some time after my abortive conversation with the Godfather. I liked the sound of his name because I thought of it as Einstein with a W. Alan W. Einstein had no doubts about the existence of gravitational waves. There was Hulse–Taylor and there was the sheer mathematics. But it was simpler than that. 'Waves are everywhere,' he said, holding his arms out wide. 'You can't move without bumping into a wave. It's not surprising there should be gravitational waves.' He had even answered an unusual undergraduate examination question on them, 25 years ago at MIT. 'An angry Massachusetts driver is waving his fist at another driver. Calculate the gravitational waves he generates.' Or something like that. The undergraduate Weinstein answered it correctly and went on to become a professor. 'It wasn't that hard,' he says now.

He looked around the lab like a parent lovingly but anxiously scrutinizing an awkward child. 'This is the really hard thing. Finding big enough waves to detect. Fists don't do it.' He smacked his fist into an open palm, thus – in a small and imperceptible way – rearranging the texture of the universe and sending off a platoon of extremely small gravitational waves into the non-existent ether. 'Unfortunately. We need a lot of mass. There just isn't enough mass down here to detect. You have to go to extraterrestrial sources. Yeah, a black hole sucking up the solar system ought to do it.' The Caltech apparatus was a pocket-sized version, almost a Dinky Toy, of the big one I had been to see in the wilds of Washington. The arms were only 40 metres long instead of 4 km, one along one side of the vast hall that used to be occupied by a synchrotron, the other poking out through the wall and disappearing under a road. But even at 1/100th scale I had a strange sense of déjà vu seeing the familiar right-angle geometry of pipes and lasers and dangling end masses and complicated contraptions for insulating it all from the outside world. I had to put on disposable boots again and a hairnet. We were in some chilly basement at Caltech, grey and metallic, vacuum-packed, sealed away from the sunny, leafy, laid-back, sociable campus above. Back in the inner sanctum. The realm of truth.

'Could you be the first to detect a wave?' I said.

'We don't have a prayer. We're a prototype.'

Other than black holes, inspirals, the Big Bang, angry Massachusetts drivers, marking papers, and the like, Alan was mainly worrying about his wife. How well did he really know her? What was she going to do next? 'I have no idea,' he reiterated. He said it with a genuine passion and conviction. It wasn't just an abstract, academic point, he really was deeply mystified by his own wife. He found her unpredictable. It wasn't rocket science, but only in the sense that had it been rocket science, he was in an ideal position (with a PhD in particle physics) to figure it out. He wished that it *was* rocket science. Mrs Weinstein was, in some way, beyond rocket science, an anomaly, irreducible to a set of equations. No mere calculus, no Riemannian geometry in 11 dimensions, was going to come up with the answers. There probably wouldn't be a question about her on the MIT exam paper – not unless she was waving her fist furiously at her husband. Which, by the sound of it, was possible.

Alan was a man who felt everything deeply. He was as sensitive as the great interferometer he had helped to engineer, as thoughtful as a seer who had roamed the wilderness for 40 days and 40 nights. Or more. He was a Jewish New Yorker who had been transplanted to the West Coast. He looked a lot like Richard Dreyfuss in *Close Encounters of the Third Kind*. He had visions of the remote past and the distant future and he grappled with reality and absurdity.

'You ever hear of an ouroboros?'

'Yes,' I said.

But Alan was rushing on and he went ahead and grabbed a pencil and drew one for me on my notebook, in case I needed to refresh my memory.

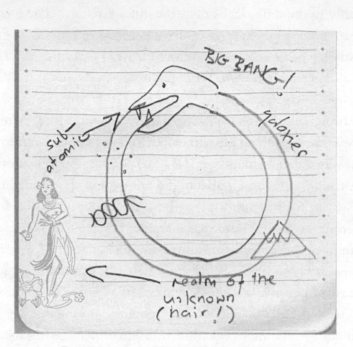

The Weinstein Ouroboros/Hawaiian notebook

A picture of a serpent eating its own tail. It was a symbolic image of the entire universe.

'See all this at the top?' he said. 'The extremely small and the extremely large. Subatomic and cosmic. Converging at the Big Bang.' He spoke breezily, dropping in a few numbers, as if this was all perfectly straightforward and uncontroversial and familiar to one and all. Then he shifted gear. It wasn't quite a sharp intake of breath, more a note somewhere between impatience and resignation. 'Down here [he tapped the mid-section of the snake with his pencil] this is all the stuff we have access to, every day.

Men and women. Mountains, viruses, even planets and galaxies. But, and this is the thing that always hits me, human beings – what do we really know about them? The "human sciences"? They're a joke. Maybe they won't be a joke a century or so from now. Maybe they can come up with some equations. But until then we are really in the dark. Does my wife still love me? Did she ever? Does anybody really love anybody? These are questions we don't honestly have answers to. We're groping in the darkness. This is the realm of the unknown. Whereas, if you think about it, all the other stuff, that seems so hard, what we're made of, where we come from, whether the planet is going to live or die, it's really easy by comparison. We can come up with an interferometer to tell us what is happening out there. Or a hadron collider. But show me the instrument that can look into the murky depths of the human heart.'

I had to admit that I didn't have one.

Alan turned everything upside down. Everything we think we know, we don't know. Everything else, all the things we don't know: this is at least knowable, in principle. Even though he had no clue about his wife, he was fairly confident about the origin of the universe. 'The Big Bang is really very simple, even though we don't understand it,' he said. 'Do you know what we say about black holes?'

'No,' I said.

'We say that "A black hole has no hair."'

'Right,' I said, as if I knew what he was talking about.

'You see, look at my hair.' He ran a tentative hand through his thatch. 'Or better still, *your* hair: it's tangled, it's thick, it's abundant, it goes this way and that, it has strands, layers.' It was true, I hadn't had a haircut for a while and my hair was running wild. I was starting to look like a mad professor, or Christopher Lloyd in *Back to the Future*. In fact I was making even the mad professors look reasonable. (It was odd, therefore, that one taxi driver had asked me flat out if I was David Beckham. I put it down to my English accent, his rear-view mirror, and the fact that there were giant billboard images of Beckham all over town, stamping themselves on his retina.)

'A black hole is not like that. You can come up with equations that define it.' Hawking had proved that you could say everything there was to say about a black hole with only three key properties: mass, angular momentum, and charge (the so-called 'No-Hair Theorem'). We didn't know black holes too well, and we were still only 90 per cent certain that they actually existed, but we knew that there wasn't that much to know about them. Alan wanted to explode the myth of complexity. Physics is the opposite of complex. Black holes and questions about the origin of the universe are really quite simple, it is everything else – like how am I going to get through the day? does my wife still love me? – that is really complex. 'We know more about atoms than we know about our wives. I'm more confident about gravitational waves than I am about her.'

We were sitting in the lab where a couple of his graduate students were working on their theses. 'We [he was talking about human beings generally] have the kind of energies that support complex structures. This is why we have disciplines like psychology and sociology.' He tapped his pencil back on the ouroboros. 'In either direction nature gets simpler and we can escape complexity. Energy densities in the early universe are too high to support complex structures. We have Einstein, we have laws of physics that work really well. So far as we have been able to explore, they still hold. But they are much much simpler than sociology. You can't apply these laws back to human society.'

I couldn't help noticing that he used the present tense to talk about events billions of years ago. Everything was history but, by the same token, history was present. Alan looked fondly on the couple of postgrads tapping away at laptops, occasionally whispering to one another, comparing notes, swapping equations, concentrating on working out the oddities of the universe. 'That's why they're doing what they're doing. That's why I'm doing it.'

All physicists were inhabitants of LIGO land: they wanted to get beyond all the noise, the racket of human existence, the shocks and the static, the incessant chatter. It was just as Dante had said, we were in the middle of the way, in a dark forest. The dark labyrinth, Galileo called it. That was the waist area of the ouroboros. Between heaven and hell. Mr and Mrs Weinstein. The trick was to get away from it all, and go beyond good and evil, the realm of freedom and indeterminacy, in the direction of ultimate simplicity.

Here is a short list of some of the things, as they occurred to me, that cosmologists – even though theoretically they were dealing with everything, the cosmos – did *not* have to deal with (so long as they remained fully focused on the job): 9/11 and suicide bombings, pollution and global warming, poverty (their own and others'), murder and crime generally, schism, disasters, rainy days, the lack of rain, disease, paedophiles, earthquakes, floods, massacres, the latest results of West Ham United football club (or substitute any other name), the latest news national and international: a long-term holiday from everything that comes under the misleading heading of 'reality'.

Being a physicist was like the fathomless laser barrels of LIGO: massively insulated against short-term shocks and tremors.

Alan reminded me a lot of Unc. 'The whole ethos of physics,' Alan said, 'is to get away from what we are.' Physicists dreamed of escaping, transcending the human condition. 'When you move away from human complexity, the level of certainty is higher. The things we know we know really well.'

'That's why I never got on with biology,' Unc agreed when I asked him about it. 'Too messy.' We were in England, walking on a beach in Dorset, wrapped up against the wind, looking out at the great turbulent oneness of the sea. Under our feet was the granulated debris of rocks, worn down and diminished over long tracts of time.

He never much got on with a lot of things: at school, literature, art, history, French. Later, his wife (and yet he must have got on with her at some stage). He had nothing in particular against foreign languages (he was good at German, for example, and once had a decent shot at Japanese) but it always turned out in the end that they were being spoken by other human beings. Whenever he went to a new place, another country, he always returned faintly disappointed, because humanity had preceded him there. Perhaps this explained why he had once seriously contemplated taking off for outer space and got as far as Houston mission control where he kept an eye on moon rockets and the like. 'I'm leaving consciousness out of it,' Unc explained. 'You don't need it to answer the equations.'

In a way, I blamed myself. I may have been personally responsible for turning him against human beings. I was the one who pushed him off the steps into the rosebushes in our front garden when he was aged around five or six. It was a kind of experiment. I was interested to know if it would hurt and Unc more or less confirmed that it did. And then, a few years later, for reasons that are completely unclear to me, I tossed his gyroscope out of the upstairs window, smashing it to pieces on the ground below. He loved that gyroscope. Much much later, still suffering a degree of remorse over my act of juvenile vandalism, I bought him a replacement, but it would never be the same.

Physicists were like monks, except they dreamed of a sinless world that was hard, compact, and shiny, with glass towers and silent sliding doors; or else infinitely small, dense, and seamless.

31

The Caltech campus, built around the turn of the century, is reminiscent of some palatial Spanish hacienda, all pillars and shady arches and fountains and cloistered calm and purple trees and buildings the colour of sand, a step away from the West Coast. No wonder that Alan Weinstein thought of the Earth as a beach with gravitational waves breaking on the shore and interferometers striving to track them back to their 'nonlinear source'.

While I was on campus it was being used as the backdrop for a TV series. As I went around talking to people or going to labs, always hoping to bump into the truth, I was always running into crowds of well-coiffeured young actors and their camera crew and their truckload of healthy snacks. Eventually, when I got back to the East Coast, I caught one of the programmes (*Numbers*, or '*Numb3rs*') purely by chance on CBS. I recognized all the leafier, more scholarly settings. The hero was a physicist or mathematician or both. His brother was a police detective. Whenever the detective, downtown, would get stuck on a murder or some other crime, his geeky brother from Caltech would get drafted in and would have to drag himself away from a bunch of particularly interesting particles or equations and focus his great mind on some sordid human interaction. The presupposition of the scriptwriters was that he would be better at working out who and why and how than his gumshoe brother, or at least have some kind of edge, by virtue of applying quantum physics to crime. Everything Alan Weinstein had told me suggested that this was the reverse of the truth. Physicists have no special edge when it comes to working out what human beings get up to.

The mystery goes beyond the feeble machinations of the most labyrinthine thriller. Even if you work out who, the question remains: why? We can list the motives: lust, lucre, loathing (the three Ls), but why even have these motives in the first place? For that we need to go back to some kind of generic theory of human nature, perhaps involving parents or grandparents or the behaviour of ancient hunter-gatherers roaming around the African savannah or, conceivably, ancient fish. But still we could only come up with a grid of possibilities. A menu of maybes. We would still be uncertain as to which particular course of action any one individual was going to adopt on any one day, whether they were going to do something wise and wonderful or, on the other hand, stupid and catastrophic. Within certain well-defined limits (it is unlikely, for example, that my friend Jonathan is about to sprout wings and fly), just about anything is possible. Every solution to every plot made about as much sense as any other. On any given day, just about anybody could have done it (whatever it was). The strange question from the flame-haired woman at the gas station in Hanford, 'Do you have a knife?', made a weird kind of sense. It was as likely that I (or anyone else) would be about to take her bag or conceivably her life as just about any other scenario. Give Alan Weinstein a dead body and ask him whodunit, and he would give a confident and robust response: 'I have no idea.' Maybe he could calculate the gravitational waves generated by a blunt instrument being smashed into someone's skull or a trigger being pulled, but he almost certainly couldn't detect them. 'We have the cool ability to be different,' he said, 'but in a soup you can't be different.' He was thinking of cosmic soup, the kind the stars, and then beings, formed out of.

Weinstein had a strong sense of the limitations of science. For example, he didn't think (it was yet another disappointment for me) that an anti-gravity device was a serious possibility: 'You can't neutralize gravity. There is no negative mass.' On the other hand, did not the very consistency and regularity of gravity mean that – and in this astrophysics would coincide with astrology – we were all driven by exactly the same forces and that, therefore, rather like billiard balls pinging off one another, our trajectories through life could be (with a big enough computer) precisely calculated?

Laplace, the great French physicist, asserted with immense optimism, at the end of the eighteenth century, that it should be possible to predict everything. To attain omniscience. All you needed to do was collect enough information about the universe at t_1 for you to be able to work out what it should look like at t_2. Laplace looked back to the orderly law-abiding universe of Newton: if everything was so totally mathematical, then a couple of savvy equations ought to pretty much wrap it all up. In this sense, astrology, predicting the course of events on Earth on the basis of observing the motion of the heavenly bodies, had it right, but it was simply an inadequate short cut: it could never manage to collect enough data and no amount of horoscopes waffling on about ascendants and houses and the disposition of Mars or Mercury was ever going to cover up the massive information shortfall. The zodiac simply misses out too much stuff, so that every horoscope is simply a wild stab in the dark. (The other major weakness, of course, is that astrology, by observing the stars, is better at looking into the past rather than the future – not to mention the fact that it had difficulty accounting for the differences between me and Unc.)

Maybe it was the fallout from relativity or quantum physics, maybe it was another couple of centuries of screwball behaviour on Earth, but most scientists have more or less given up on the deterministic model of the world. Here on Earth chaos rules. 'We understand plasma,' as Alan Weinstein said. But for the physicist, human beings are inscrutable, baffling, fundamentally unfathomable. You would be better off asking a postman or a butcher or a car salesman – or the woman in the gas station – for their opinion about who killed Roger Ackroyd. Or whether or not they understand their wives – or their husbands.

And yet: the origin, the first moment, the primeval atom, the Aleph – it seemed to hold out the prospect of at last understanding everything. Was everything, after all, not contained within its infinite density? All those microscopic gravitational waves, billions of years old, must have, somewhere among them, one with our name on. I think this was what Chateaubriand meant – writing at the beginning of the nineteenth century in his *Génie du christianisme* – when he wrote that the entire universe was created in one fell swoop and was then exactly as it is now, young and old all at once. Everything, birds and their fledglings, sheep and their lambs, ancient things, ruins, fossils, eroded headlands, trees bent over the abyss, Albert Einstein, Alan Weinstein, they were all there from the very beginning (whenever that was). There was no golden age of unwrinkled innocence. Everything which is now has always been. As it was in the beginning, is now, and ever shall be.

32

I was sitting outside the Red Door Café on the Caltech campus. I was killing an hour, drinking a latte and reading the *Los Angeles Times*. Students were roaming about or lying on the grass talking and laughing. A couple of guys at the neighbouring table were bouncing some abstruse equations back and forth, like it was ping-pong. I was shaded from the direct light of the sun by some kind of straw parasol over my head. The blue jacaranda trees swayed elegantly. Suddenly, as if in a mirage, everything in my field of vision started to shimmer, very faintly, so faintly as to be semi-imperceptible. The campus was starting to breathe in and out, throbbing, pulsating, on a microscopic scale. Perhaps it would be truer to say that I was feeling or intuiting the vibrations rather than seeing them. Or that, in some odd way, the vibrations were passing through me, and that I was projecting them out on the space-time continuum at large. Gravitational waves were washing over me, compressing and relaxing me, like a Swedish massage. They had been doing it all the way along, but I'd only just noticed. It was obvious, I thought to myself. How could I have missed it? The people around me, the trees, the massive buildings, shyly, subtly, very tentatively, they were all dancing, shimmying in an invisible belly dance. I wanted to sing out and let everyone else around me know, but I didn't dare move for fear of losing the pulse.

Then, as suddenly as it started, the shimmering faded away, and I was left with the world as it had always been, rock-solid, impermeable, unquivering.

I was left thinking of Albert Camus. Back in 1946, after the Second World War, the French novelist and thinker had once made a visit to an American campus. He was impressed by the women at Vassar, lazing on the lawn. But the moment I had in mind is recorded in one of the school exercise books he used as a diary. Camus suffered from chronic tuberculosis and spent a lot of time in bed as a youth. Lying there in a sunny room, he came to form a degree of intimacy with inanimate things, things that it would be hard in fact to describe as things. He seems to have had the ability to renounce his own ego, in a quasi-Buddhist way, and identify fully with objects outside himself. In his *Notebooks* he is often lying in bed, in Algiers, staring out of the window at the sky, taking note of the passage of clouds, the play of light, the colours of the sky. Looking through the window, like Galileo looking through his telescope, like Einstein conducting his thought experiment, Camus is riding the light – and feels released from carnal being and poverty:

> A flash of light fills me with a confused and bewildering joy. What am I and what can I do other than enter into the patterns of the foliage and the light? To be this ray of light into which the smoke from my cigarette vanishes, this sweetness and this discreet passion pulsating in the air.

It may seem clinically inadvisable for someone suffering from a major lung condition to be smoking as much as Camus does. But there is clearly something about the smoke – beyond mere inhalation – its weightlessness, its insubstantiality, which fascinates him. It provides a kind of passageway – a stairway – if not to heaven then at least to the condition of light. Camus feels as if he is going up in smoke along with the cigarette. He feels, obscurely, that he is blowing himself out into the air and becoming light, dissolving into sky, air, water. 'If I attempt to reach myself,' he continues, 'it is in the depths of this light.' Camus identifies with what can be broadly defined as cosmic phenomena. Camus could feel the waves, the vibe.

On his journey to the United States, he was deeply impressed by New York, developed a taste for ice cream, fell in love a few times, contracted a fever, listened to jazz in a Greenwich Village basement bar, and finally took ship again for South America. He went up on deck at night, probably lit a cigarette, and pulled out his notebook again:

> Marvellous night on the Atlantic. This hour when the sun has disappeared and the moon has just barely been born, when the west is still luminous and the east is already dark. Yes, I've loved the sea very much – this calm immensity – these wakes folded under wakes – these liquid routes. I've always been torn between my appetite for people, the vanity and the agitation, and the desire to make myself the equal of these seas of forgetfulness, these unlimited silences that are like the enchantment of death. I have a taste for worldly vanities, other people, faces, but, out of step with this century, I have an example in myself which is the sea and anything in this world that resembles it. O sweetness of nights where all the stars sway and slide above the masts, and this silence in myself, this silence which finally frees me from everything.

Stars, space, silence, the swirling patterns in the wake of a ship: Camus seems to be sailing away from the Earth, riding the waves like the Silver Surfer. Like the Weinsteinian physicist, like Unc. The psyche fading away into the Brownian motion of liquid and photons, a kind of discarnate gas or vapour, dispersed out into vacant interstellar spaces, merging with the immaterial and oblivion, effortlessly transcending the illusion of individuality. Even before dying, Camus saw himself translated into the form of light.

Jean-Paul Sartre, the philosopher, was his greatest adversary. They first met in Paris in 1943, in the midst of war, and became friends and comrades. But even before they became enemies, there was always an antagonism between them. There was jealousy too: Sartre looked like something hanging off the outside of Notre Dame cathedral whereas Camus – and Sartre could never quite forgive him for it – had movie-star good looks, he was a French Bogart or James Dean. But the key difference between them was that Sartre did not believe he would ever turn into pure light. To be like a thing, like light, as Camus imagines, is like trying to be God, Sartre says (what he memorably called 'the-in-itself-for-itself'). Can't be done. We are (as Woody Allen said) at two with nature. And we can't empathize with our fellow humans either: 'Hell is other people', war is natural, and love is just a hoax or a 'metaphysical impossibility'.

Sartre thought of the beginning in a similar way. In his great existential novel *Nausea*, he sings a kind of love song to the very idea of beginning:

First of all the beginnings would have had to be real beginnings. Alas! Now I can see so clearly what I wanted. Real beginnings, appearing like a fanfare of trumpets, like the first notes of a jazz tune, abruptly, cutting boredom short, strengthening duration; evenings among those evenings of which you later say: 'I was out walking, it was an evening in May.' You are walking along, the moon has just risen, you feel idle, vacant, a little empty. And then all of a sudden you think: 'Something has happened'. It might be anything: a slight crackling sound in the shadows, a fleeting silhouette crossing the street. But this slight event isn't like the others: straight away you see that it is the predecessor of a great form whose outlines are lost in the mist and you tell yourself too: 'Something is beginning'.

All beginnings are echoes, distant relatives, of the origin. But Sartre refused to give much credence to even such minute, unpromising, trivial beginnings as these. The beginning, any beginning, belongs to the sphere of the 'adventure', to narrative, not to real life. Just as Stephen Hawking suggested that maybe there were no 'boundary conditions' in the 'history of time', that everything flows without a great drama to wind up or shut down, so too, in everyday life, according to Sartre, 'nothing happens. The settings change, people come in and go out, that's all. There are never any beginnings. Days are tacked on to days without rhyme or reason, it is an endless, monotonous addition . . . There isn't any end either . . . everything is like everything else.' So every true-life story is always a lie. Sartre argues that 'you have to choose: to live or to recount'.

Camus died in a car crash on the road to Paris in the first days of 1960. He had written love letters to at least three women on two separate continents before setting out. But he was also carrying with him the manuscript to a novel that he had started but that would remain for ever unfinished. *The First Man* recounted his origins in Algeria, his mother, his education, his first loves, swimming in the Mediterranean, all bathed in the glow of nostalgia. To Camus everything was a new beginning and therefore meaningful. To Sartre, the denier of beginnings, everything was meaningless and futile and monotonous. Probably all cosmologists would agree with Sartre (even if they would choose different words) that we live in a 'detotalized totality'. That is what the universe is like. The question is, which part of that phrase do you put the emphasis on: the *totality* or the *detotalized*?

Sartre would say that nostalgia is stupid or absurd. But I suspect that cosmologists would come down more on Camus's side than Sartre's. While writers like Sartre refuse to take the idea of beginnings and endings seriously any more, or denounce it as a literary conspiracy, or joke, scientists have begun to take seriously the idea of a great overarching narrative that would fit everything into place. In the nineteenth century, positivism honoured science for ignoring any questions to do with origins and destiny. That, Auguste Comte said, was the business of theologians and teleologists, the stuff of myth and metaphysics, not physics, and good positivists should really be concentrating on the bit in the middle, the only part that can be shown, investigated, recorded. It was the end of the beginning, to Comte's way of thinking. But in the twentieth century, physics moved into and took over metaphysics. Physicists really want to begin at the beginning and go on till the end and then stop. The earlier term for physics and science more generally, 'natural philosophy', seems more relevant now than ever. The beginning, which began as the preserve of prophets, and was later derided and despised, never really went away. It was always hidden away in the corner of mirrors, at the far end of laser tunnels.

Without realizing it, we are all cosmologists, more or less, dreaming of where and how it all began. If cosmologists agree on anything, it is that the universe is expanding and if you rewind the tape you end up – or rather start off – with a singularity: a single point of origin, followed by the mother of all explosions. Philosophically and politically, we are left with two serious options in the way we look at the world. Implicitly harking back to the primeval atom, we can see that something extremely unified has been lost in the course of our development, but that we can try very hard to get it back again. This was Camus, giving voice to O-theory.

Giving pre-eminence to sameness. On the other hand, following the arrow of time forwards in its trajectory, we can embrace the joy of fragmentation and insist on the sheer divergence to which every individual entity in the universe owes its existence. So runs X-theory, the opposite of coherence, with everything splitting away, diverging, from the centre, with everything and everyone irrevocably split and sundered. The realm of difference. In other words, we can choose to give priority to the primordial state of the one or we can give up any notion of unity as a nostalgic illusion and put all our faith in the many and sheer multiplicity. Sameness or difference, you choose.

When John Lennon, looking ahead to the future in 'Imagine', imagined that 'the world will be as one', he was being a unificationist. On a more nostalgic, yearning note, Paul McCartney's 'Yesterday' attributes the lost unity to the past. Both are O-men in their sympathies. But whatever their songs say, the fact remains that Lennon and McCartney and the Beatles split up. It is ironic that the Rolling Stones, a band of X-men, who sounded a much more divisive note, giving expression to the need to escape from cloying dependency, and stated in 'Ruby Tuesday' that 'Yesterday don't matter when it's gone', succeeded in sticking together, by and large, the odd drug-induced fatality aside.

It was on that afternoon, hallucinating in the sun outside the Red Door Café, having my Joe Weber moment, that I realized that there is a Sartrian side to Einstein and a Camusian side. He is part Beatles, part Rolling Stones, McCartney and Jagger superposed. He developed an X-theory, full of alienated frames of reference, with people marooned on their own separate planets with asynchronous clocks, and yearned to pull them all back together again in a state of harmony, a single collective grand unified theory. If you could only detect gravitational waves from the beginning of time, what I thought of as the soul of the universe, it seemed to me that there was hope for O-theory after all.

33

Considering that Caltech was such a graceful, leisurely sort of place, shaded by palm trees, their fronds wafting amid sultry breezes, I seemed to be doing a lot of worrying. Right now I was worrying about the Big Bang. Science hadn't really conquered anxiety, not yet anyway (all it could do was remind you of that blissful time when anxiety did not exist). It was something Kip Thorne had said in the middle of our brief non-meeting of minds, and I should have picked up more on it then, but it kept coming back to me now, as I roamed around campus, getting buffeted by passing waves.

'I don't know if the Big Bang is a limit point or not,' he said. He had a grey beard. It was a little on the short side if he really wanted to look like God though. I have a soft spot for the casual admission of ignorance, but this was something I didn't think we could afford to be too hazy about. We were talking about the beginning of everything, after all. 'Big Bang theory is in a primitive state,' he went on. 'We don't understand the quantum physics well enough. From 10^{-34} seconds on in, the Big Bang is speculative. It's not hard science.'

That was the merest fraction of a nanosecond after the emergence of time and space. A mere 0.00000000000000000 00000000000000001 seconds. After that, it was fairly plain sailing. Good enough, you might say. But, to be honest, the more I thought about it the more disappointing this was. In the early history of the universe, a trillionth of a second can be a long time. The Big Bang was one of the most solid, nailed-on theories that astrophysics had come up with in the whole of the twentieth century and now, in the early years of the twenty-first, we were admitting that we didn't really know what we were talking about. This is the best story we have and it's full of holes. In fact it's one mighty big hole through and through. That is the beauty of it. The Big Bang is the ultimate everything-and-nothing story. It attempts to answer the fundamental question, 'Why do we have something rather than nothing?' And it replies, more or less as Unc did, with his mind-bending quantum logic: we have something *because* we have nothing. A fluctuation in the vacuum. The Big Bang as someone once said is the ultimate free lunch. It offers not just something for nothing, but everything.

Humans seem always to have felt compelled to tell stories about where they come from. In China everything began with an egg, floating on a formless sea. The Polynesians speak of the cosmic coconut out of which the universe sprang. Greek myths mention Chaos, Gaia, and Eros. Australian aboriginals tell of the Dreaming or Dreamtime in which ancestral spirits summoned up life, leaving footprints behind them in the form of 'songlines'. Some say the world is carried around on the back of a turtle, an elephant or a tiger. Claude Lévi-Strauss, the anthropologist, says that what all Amerindian myths share is that they refer back to a time when there was no distinction between humans and animals and the creator of the world could as easily be a silver fox or a snake or a kangaroo as anything else. Originally these traditional tales were exclusively oral, and they long predate the literature in which they eventually appear. Stories about the origin go back as far as stories go. But all these narratives of gods, eggs, sun, sky, birds, Earth, serpents, songs, turtles and humans leave the chronology of events rather hazy.

In 1642, with this problem in mind, Archbishop Ussher of Armagh tried to give the book of Genesis a bit more precision. On the basis of a thorough reading of all the ancient texts, in the original Hebrew, the Jewish calendar, all those 'begat's, and Kepler's astronomical tables, he calculated that God created the heavens and the Earth in 4004 BC. He got started on the evening of the day preceding Sunday 23 October (in the Julian calendar), just a couple of days before the autumn equinox. Ussher's date came to be regularly quoted in annotated editions of the King James Bible, which is where I first came across it (I can remember being tremendously impressed, around the age of 10, by how clearcut it was, nothing fuzzy or speculative there at all; and I still wanted it to be clearcut). Others, using similar methods, placed the Genesis moment in 3929 or 3952 BC. They all had good evidence and reasons for their estimates. And then, in the background, there was the clearly stated divine timeframe set out by one of the apostles, 'one day is with the Lord as a thousand years, and a thousand years as one day' (2 Peter 3:8). This tended to suggest, using the days of Creation as a template, that we had been granted a grand total of six thousand years on Earth, four before the birth of Christ and two afterwards (leaving the one day of rest to correspond to the rest of eternity). Around the end of 1999, a lot of people still reckoned that the world was due to come to an end fairly shortly (others maintained that it was just computers that were doomed). The archbishop's biblical chronology was a serious, sober, scholarly, scrupulously researched assessment. But it was based strictly on an intertextual system, in which one inherited genealogy hooked into another, providing a solid, cohesive history of the world, with no gaps between Adam and Ussher, passing seamlessly from the Creation to the Flood to the

Exodus to the birth of Jesus. Archbishop Ussher and his contemporaries owed their allegiance to the tradition of telling rather than showing, with one teller telling another, who tells the next, and so on, all of whom derived authority from sheer repetition.

But there was an alternative tradition of showing, one that showed little respect for pure telling. Galileo was persecuted for his telescopic view of the heavens and forced to recant; Giordano Bruno, later in the century, was put on trial, condemned, stripped naked, hung upside down, and burned alive for his heretical beliefs, notably that there must be extraterrestrials out there somewhere. But, the eccentricities of Earth aside, the solar system and beyond, through Newton and Laplace, once it was well understood, appeared to be a stable, consistent, orderly state of affairs. It would look broadly the same tomorrow as it had looked yesterday. There was no particular logic for a beginning: Genesis would have occurred at the time of God's choosing. The age of the universe and its very existence appeared arbitrary. The British philosopher Bertrand Russell in the twentieth century followed the same thought to its logical conclusion: the universe could have been invented five minutes ago, or yesterday, or last Thursday, with all our apparent memories of the past implanted in our heads.

> There is no logical impossibility in the hypothesis that the world sprang into being five minutes ago, exactly as it then was, with a population that "remembered" a wholly unreal past. There is no logically necessary connection between events at different times; therefore nothing that is happening now or will happen in the future can disprove the hypothesis that the world began five minutes ago. (*Analysis of Mind*)

The idea was not in the least plausible but it was entirely unfalsifiable, no amount of argument or evidence could shake its tenuous foundations, like a thin but earthquake-proof dwelling.

We had geology, we had fossils, we had elongated the chronology. But it was not until Einstein that we had an equation that suggested that the universe could be expanding and must therefore have had an origin in time and space. The idea was so contrary to the prevailing Newtonian concept of the stable universe that even Einstein rejected it. In fact he was concerned that, on his calculations, with so much gravity about, there was a strong chance that space-time must be shrinking and folding in on itself. Newton had worried over the problem and proposed that everything neatly cancelled out to preserve the status quo. Einstein came up with what he called a 'cosmological constant' (Λ, or lambda) that balanced the equation, thus – as the medieval theologians used to say – 'saving the appearances' and keeping things nice and steady. But the equations were there: people like the obscure meteorologist Alexander Friedmann in Russia, in 1922, could infer expansion and thus a chronology from them, 'time-varying solutions'. Edward Said, the Palestinian literary critic, once said (in his book called *Beginnings*) that the notion of an 'origin' is strictly divine and intransitive and that at some stage we shifted to the idea of more human, secular, intentional, transitive 'beginnings'; and that some time in the course of the nineteenth and twentieth centuries, moreover, we lost our faith in any kind of beginning and more or less gave it up altogether.

The Belgian priest and cosmologist Georges Lemaître, who studied at Cambridge under Eddington, saw no serious conflict between science and religion. The Big Bang, the great singularity in the sky, had a reassuringly monotheistic sound to it. In 1927, working independently of Friedmann, Lemaître traced the implications of relativity theory backwards to assert that the universe must have begun with the disintegration of a 'Primeval Atom', a moment of creation, 'a day without yesterday' that was perfectly compatible with the book of Genesis. In the twentieth century the beginning became the focus of cosmological thought.

The situation was roughly the reverse of the one described by Said. For the first time the origin acquired scientific respectability. Lemaître summarized the new mentality: 'The evolution of the universe can be likened to a display of fireworks that has just ended: some few wisps, ashes and smoke. Standing on a well-cooled cinder, we see the fading of the suns, and try to recall the vanished brilliance of the origins of the worlds.'

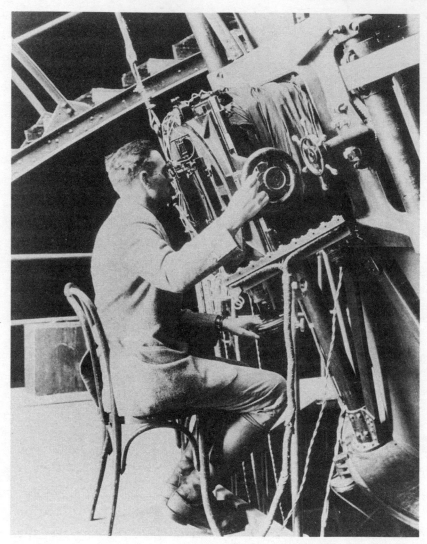

Hubble at work (strong resemblance to Unc)

34

The Friedmann/Lemaître theory, occasionally doused by Einstein's own scepticism, remained speculative until 1929 when Edwin Hubble came up with experimental evidence. He wasn't even looking for it. Hubble had read Jules Verne as a child – *From the Earth to the Moon* had been a favourite – and had been inspired to take up astronomy. For a while (as a Rhodes scholar) he studied in Oxford. When I looked at pictures of him, especially when seen from the back with his head plugged into a giant telescope, he reminded me a lot of Unc. He was tall, a boxer and a basketball player. To tremendous acclaim, Hubble had already established (using 'Cepheid variables') that there were galaxies – 'island universes' – beyond the Milky Way. As he worked on determining their distance, he couldn't help noticing that the further away stars were the more they were marked by a 'red shift'. The spectrum of light – what Hubble called 'emission nebulosity' – would be a crucial factor in establishing the structure and history of the universe.

It turned out that Unc knew Auguste Comte's 1835 *Cours de*

philosophie positive. I was in Oxford one time, looking out of the telescope he kept in his back room (it was day and I was idly peering into neighbours' houses and watching the pigeons). I had to admit to some surprise. He didn't usually seem that inter-ested in the works of the French founder of positivism. But the passage that Unc quoted at me (pulling it out of a textbook) was specifically concerned with stars. Comte said that there was no way we could ever work out what was in stars. It was just beyond our powers.

> On the subject of stars, all investigations which are not ultimately reducible to simple visual observations are . . . necessarily denied to us. We can conceive of the possibility of determining their shapes, their distances, their bulk, and their motions, but we can never know anything of their chemical or mineralogical structure. Our knowledge concerning their gaseous envelopes is necessarily limited to their existence, size, and refractive power.

'A reasonable enough proposition,' said Unc, leaving a pregnant pause. 'On the face of it.'

'You mean wrong?'

'So wrong it's not even wrong any more,' he replied enigmatically.

Which explains why astronomers love to quote Comte. It shows they are smarter than French philosophers (something they never really doubt anyway). True, stars are up there and we are down here. Nothing is further away than stars. And yet the quality of light emitted by stars turns out to provide the key to understanding their chemical composition (such as helium), temperature, and pressure – the basis of spectroscopy. Hubble understood that the colour of light provides the crucial clue to their motion too.

Unc didn't have to explain the Doppler effect to me. I could hear it any day of the week. When a train blows its horn as it speeds by, the sound rises in pitch as it approaches, and lowers as it recedes. Similarly the whining sound of a Formula One car first rises into an unbearable scream as it flashes by and then falls away ('weeee-oooow', roughly). An ambulance siren: same. The sound waves bunch up and compress to produce higher frequency and higher pitch, and stretch out into the distance to produce lower frequency and thus lower pitch. But the Doppler effect (first defined by Christian Doppler in 1842) also applies to stars. Others (such as Huggins and Slipher) had noticed the red-shift effect in distant stars: a shift in the spectrum of a star relative to our own sun (or 'absolute luminosity criteria'). You can actually see it in a spectroscopic analysis: the lines, indicating wavelengths, are shifted to the right of the strip (towards the red). A few showed a blue shift (towards the left). Now, red light has a long wavelength (around 700 nanometres), blue light a shorter wavelength (400). That meant that the red-shifted stars, radiating light with a longer, stretched-out wavelength, must be flying away from us, like a passing train, like the ambulance speeding into the distance. Hubble, using the 100-inch Hooker telescope on Mount Wilson, just outside Pasadena, then the biggest in the world (remaining so until 1948), set about tackling the mystery of the red shift.

I could actually see the top of Mount Wilson, porcupined with antennae and domes, from the Caltech campus, so I hired a car from Alamo and drove east, through peach and orange orchards, snaking up the Crest Highway into the San Gabriel Mountains. Mount Wilson is only slightly less than 5,710 feet high, less than half Mauna Kea. The road was steep and winding and lined with fir trees, but I didn't have to acclimatize, or take puffs of oxygen, and I didn't feel drunk either. It wasn't even particularly chilly. The antennae I could see turned out to be a bunch of aerials for television and radio broadcasting (one of the Caltech physicists, Sterl Phinney, later told me he had calculated 'the flux of radio power' in the middle of the 'antenna farm' as 5 watts/m^2). The observatory itself is almost like a log cabin in the hills, except that it is an observatory, with the shining white dome of Hubble's telescope poking up out of the pines. The single-storey wooden astronomers' dormitory is known as 'the Monastery'. I could look down and see Los Angeles spread out below me through the haze. To the west, barely visible, the Pacific Ocean with the Hawaiian islands 2,000 miles further away. It must have seemed natural to hop from here to Mauna Kea. Michelson had carried out some of his light-speed-testing experiments up here, not to mention measuring the diameter of Betelgeuse. If you could ignore the curvature of the Earth, light would take all of a twentieth of a second to go to Hawaii and back.

Few clouds, plenty of light, even flashes of lightning in the distance over the top of purple hills, but no stars, not one, not unless you count the sun. But when I was on Mount Wilson they had the Hooker mirror out for cleaning and recoating and display, sitting on a plinth. One hundred inches across, that's more than 8 feet, taller than a basketball player and a lot wider. The largest solid glass mirror ever made (subsequent mirrors of that size were honeycombed in the back). John D. Hooker was a local businessman who put up the funding for the mirror. The First World War was looming when this mirror was being made. It saw first light on 2 November 1917, while – far to the east – several armies were encamped on the plains of Flanders, thousands were dying in the mud every day at Ypres and Passchendaele, clouds of mustard gas blocked out the sun, and fighting was raging on the Mesopotamian Front. The mirror room was fireproof and earthquake-proof. The whole purpose of the mirror, or so it seemed to me, was to focus on something other than the horrors of the trenches in Europe. Put your eye to the lens: it was not just that all the glories of the heavens suddenly became visible, it was also that all terrestrial struggle was eclipsed and faded from sight. No more corpses, massacres, rats, machine-gun fire, terror and pity, nothing but light. Back then, before the expansion of Los Angeles and the inexorable rise of Hollywood, the air must have been clear and purer than in the surrounding valleys. The mountain was a place of coyotes, eagles and bears. No roads. You needed a burro to get up there. The original Wilson (Benjamin) was a hardy pioneer and cowboy.

The mirror was actually manufactured in France, at the French Glass Works Factory at St Gobain. Dark and thick and green, it looks like the bottom of a particularly generous bottle of wine. St Gobain, which had cast the earlier 60-incher ten years before, was the pre-eminent mirror-maker in the world at that time, but even they didn't have the means to pour 9,000 pounds of glass at one fell swoop. They had to do it in three takes and bubbles of air were trapped between the layers after every pour, so it looks as though someone or something is breathing down there, trapped in time, and the bubbles of air aren't quite making it to the surface. Seen from the side, the waves and currents of the ocean appear to be frozen in there too, fixed, like a photograph. It was so flawed that the glass was originally rejected as useless. Even Hooker despaired of it. The mounting, the rails, the housing, the clock drive, five years alone for grinding and polishing, shipping it all in through the Panama Canal, lugging the lot up the mountain, building roads to lug it up, not to mention delays caused by the war effort and at least one last-minute telephoned threat to blow it up: all in all getting the telescope built and up and running took over a decade.

To celebrate, the founder and director of Mount Wilson, George Ellery Hale, invited the English poet Alfred Noyes, then at Princeton, to the observatory to record the grand opening – the launch – in poetic form. He wrote the epic 'Watchers of the Sky', a hymn to astronomers:

> The explorers of the sky, the pioneers
> Of science, now made ready to attack
> That darkness once again, and win new worlds.

Noyes couldn't help but think of the war too and saw the telescope as an alternative, perhaps a solution, like a prayer, flying up into the light.

> The noblest weapon ever made by man.
> War had delayed them. They had been drawn away
> Designing darker weapons. But no gun
> Could outrange this.

When all the hurrahs and the speech-making and poetry were over and someone finally looked through the eyepiece, it was thought that the whole exercise had been a disaster. Several overlaid blurry images of Jupiter presented themselves. It was sub-Galileo. Eleven years of construction down the drain. This very expensive French glass was pronounced worthless. Then it transpired that some workmen had left the dome open during the day and allowed the glass to heat up. It wasn't until 3 a.m. when the glass had cooled that a sharp image of a distant star was obtained and the doubters were finally routed.

In 1928, Hubble's assistant, the former janitor Milton Humason, measured the Doppler shifts of far-flung galaxies while Hubble worked out the distances. Putting the two together, they established that the furthest stars were showing a higher red shift than those that were closer. Systematically and consistently. Not only were they receding, but the further away they were the faster they were receding. It was a pure linear relationship.

'He could actually draw a straight line through the data,' Unc said. Unc was seeing the graph in his head, with velocity on the y-axis and distance on the x-axis, and all the dots, corresponding to stars, neatly corresponding to an upward graph, heading up cleanly at 45 degrees. In fact the data wasn't really that neat, but you had to allow for errors.

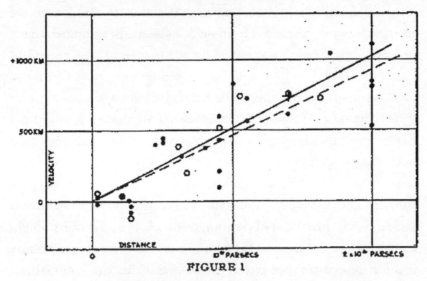

From Hubble's paper, the graph relating velocity to distance; Unc's 'straight line'

Hubble's 1929 paper 'A Relation between Distance and Radial Velocity Among Extra-Galactic Nebulae' effectively introduced Hubble's Law, which would become the keystone of modern cosmology: recession velocity is proportional to distance (v = Hd where H is the Hubble constant). In short, further = faster. Twice as far away, twice as fast. Therefore, once upon a time, throwing the process into reverse, everything must have been packed tightly together – in a state of infinite compression – and then all been blown outwards. As if in an immense primordial explosion. Hubble's observations were the first to bear out the Lemaître theory of an expanding universe in which distance is also a measure of history. And the idea that there had once been a creation event, a time zero. It was just a question of tracing the graph backwards towards the intersection of the x and y axes, 0,0.

Einstein had got to Mount Wilson before me. He eventually conceded that, in assuming equilibrium, he had made his 'biggest blunder' and on 29 January 1931, while he was a visiting scholar at Caltech, drove up to the observatory to meet Hubble and congratulate him on rectifying the mistake. It was an act of repentance and benediction. In pride of place in the pink palace of the Mount Wilson museum, there is a photograph of the pair of them standing together with Hubble towering over Einstein. Humason was able to show Einstein the galaxies' spectra with their characteristic red shift. On 3 February Einstein gave a press conference at Mount Wilson publicly renouncing his static cosmology and endorsing the expanding universe. Later, in 1933, after listening to Lemaître speak at a conference in Pasadena, Einstein is supposed to have said, 'This is the most beautiful and satisfactory explanation of creation to which I have ever listened.'

The funny thing was that Hubble never commented on the implications of his own work. It was one of the most significant discoveries of all time. He could have become a cosmologist, a philosopher, but he preferred to remain an astronomer, confining himself austerely to observation. He seemed to feel that he had already said more than enough.

Ironically, the phrase 'this Big Bang idea' was a sarcastic put-down and came from the opposing camp in Cambridge, England. Fred Hoyle, who was then director of the Institute of Astronomy, came up with it first on BBC radio back in 1950 and he didn't believe it and wanted to discredit the whole crazy idea. Hoyle was referring to Lemaître's vision of the origin: 'In the beginning of everything we had fireworks of unimaginable beauty. Then there was an explosion followed by the filling of the heavens with smoke. We come too late to do more than visualize the splendour of creation's birthday!' He probably objected to the fact that the Pope approved of Lemaître's theory and that the priest-physicist had been elected to the Pontifical Academy of Science. But Hoyle was a visionary and science-fiction writer with inspired ideas of his own – he proposed, for example, that the solar system must have been seeded by stardust (or possibly benevolent aliens) – but he was a strict 'steady state' theorist, who reasoned that new matter was being created all the time in gas clouds between galaxies, thus accounting for the apparent expansion of the universe. But all the evidence pointed towards the Big Bang, every new observation that came along seemed to vindicate it. The primeval atomist tendency liked the label and adopted it to spite Hoyle.

In 1964 Arno Penzias and Robert Wilson, a couple of young postgrads working at the Bell Labs in Holmdel, New Jersey, were supposed to be measuring the intensity of Cassiopeia A (a supernova remnant). Their 20-feet horn reflector radio antenna was supposed to be super-low-noise. But still there was this mysterious signal, a background hiss, they couldn't get out of the experiment. At first they blamed the pigeons for perching on and crapping on the horn. 'White dielectric material' as they called it. They had to eliminate a lot of pigeons (using the 'Hav-a-Heart' pigeon trap) before they realized it was the whole universe which had been blaring out signals about its own early history from billions of miles away. In fact, they still could hardly believe it. 'Measurement of Excess Antenna Temperature at 4080 Megacycles', that was the title of their paper. You could never accuse them of hype. A few miles down the road, at Princeton, Robert Dicke had already worked out the theory, but still he felt scooped that he hadn't detected the signal first. Although in a way probably he had, he just didn't know it then. Cosmic microwave background radiation: every time you turn on the TV or the radio and get a lot of snow or static, like the sound of the surf – that's an echo of the Big Bang right there. We're living amid the dying embers of the original fireball. Nearly three degrees above absolute zero. It is chilly, but it's still measurable. It's all around us, we just choose to ignore it most of the time or blame it on pigeons. Then there was the great COBE satellite, which showed up the ripples in the background radiation that would ultimately distil into galaxies. And then there was the Hubble telescope, orbiting in space, and its terrestrial partners like the Keck that made everything visible, put it all on show. For the first time you could look billions of years into the past and actually see whole galaxies taking shape or great suns blowing up – supernovae – and producing more building blocks for the future. It was the ultimate tourist slideshow. You could show photographs of the giant gas clouds of Sirius XI. You could send postcards home of the otherwise obscure irregular galaxy NGC1427A.

The Big Bang was just a modest extrapolation from all that. Age of universe = 1/H where H is the Hubble constant. Instead of 4004 BC you ended up with a figure of 13.7 billion BC, give or take a few thousand.

Alan Weinstein said that physics hated complexity. It wanted a universe with no hair and everything plain and square and straightforward and reducible to a neat set of equations with no wifely emotions. He reckoned that he could make coherent sense of everything all the way down to around 10^{-30} seconds (what he helpfully defined as 'a billionth of a billionth of a billionth of a second') after time zero, when 'all matter and energy occupied the core of an atom' and that was the entire length and breadth of the observable universe, no more and no less.

There was only one irremovable hair stuck in the mouth of the whole story, and that was the Big Bang itself. The most singular of singularities. It was the ultimate paradox: infinite density and mass in a space with zero volume (or 'infinitely curved' space-time). You could say it mathematically, but there was no way anyone could picture it or make sense of the idea (which is probably why Einstein initially revolted against it). The word 'singularity' was just an admission that it was incomprehensible and beyond argument. It was another way of saying that the idea of an origin, time zero, nothingness, many more than one world condensed to a lot less than a grain of sand, resists analysis. Everything = nothing. 1 = 0. The Big Bang is a Zen koan posing as physics: the sound of one hand clapping. Alan Weinstein said it is 'very very simple' (and 'much much simpler than sociology'). But living as we do in the social realm of complexity, it is impossible for us to grasp absolute simplicity.

On the other hand, Kip Thorne sounded at least one upbeat note. 'I'm confident that in the next decade or two, one or other of our systems [he was talking about LIGO and LISA] is going to make the Big Bang into science.' Grav waves would come to the rescue and straighten everything out, finally, once and for all. End of story.

35

There was at least one plausible alternative to the Big Bang. Avis, back on the East Coast – still deeply sceptical about my quest – e-mailed me the latest. She was determined to shoot me down.

In the 'Big Bounce' there is no free lunch any more. For one thing to begin something else has to end. On the other hand, we are looking (as Edward Said recommended) at multiple beginnings rather than a singular origin, or at least at a serial beginner. Avis sent me a ton of mathematics (with credits to Martin Bojowald of Penn State), based on 'Loop Quantum Gravity', because it was clear I wasn't going to see this out of any telescope, not the Keck, not Mount Wilson, and not even LIGO. So I did what Einstein would have done: I tried to picture it. In fact since I was up in the hills overlooking Pasadena, not that far from Hollywood, I saw it as a movie (I had reverted to my old electromagnetic prejudices). As I looked out over the abyss, I could see it on the big screen or the YouTube in my head, but I wasn't sure if I was seeing it forwards or backwards (but then part of the point of the Big Bounce is that it doesn't matter any more:

fast forward, rewind, it's all one). I knew that what I was seeing was an immensely compressed version of the maths, as if the timer was on speed, with awesome Industrial Light and Magic special effects. The ultimate disaster movie, but with a Hollywood ending.

It was like the experiments Unc and I used to try out to do with how many times you can fold a piece of paper. They say it's a maximum of eight times. Unc and I used to get up to around ten using a very large thin sheet of tissue paper and ironing it down. It was still not enough and the small rectangle we were left with was still impossibly baggy. Except this time we were starting – looking down from the hills – by rolling up the whole of Los Angeles, and then the rest of the US, then the world, and beyond. In the Big Bounce the universe (or it might be better to say '*a* universe') pulls off the impossible trick of folding itself in two time after time, a thousand times over, then some more.

You and I, everything you know, everyone we know or ever have known, anyone, anything that ever existed, we/they are all being neatly squared away, put back in the box. All the air is being let out of the balloon. No more cities, no more people, no more streets or cathedrals or harbours, no more trees, no more land, no more ocean. A whole world evacuated. No more planets, they all dissolve into suns. No more suns, they all telescope back into galaxies. No more galaxies either, for they too are contracting and fusing into a homogeneous cloud of light. No more clouds, no more light, only darkness visible, streaky, murky, miasmic, gradually consuming itself, slimming, condensing, collapsing inwards, squeezing all that mass into a tighter and tighter volume of space, and then halving it all again, negative power on top of negative power, dividing, reducing, subtracting, annihilating, and yet losing nothing in the process. Nothing is thrown away, everything is saved, but infinitely retracted, compressed and convergent. Every difference dissolves as the oneness rushes in on itself and eats its own tail. The entire universe written on the back of a postage stamp. And then the stamp folds itself up and vanishes. At last there is an invisible circumference, only intelligible through the form of gravitational waves, and the circumference itself shrinks to a point, wave upon wave retreating and returning to the original source of pure shapeless insubstantial energy from which they came. The primeval atom is a bubble and then the bubble bursts, inwardly, imploding. All the heavens and the earths are being uncreated, cremated, deflated, cancelled, voided, erased, consumed. The universe is being born – backwards. Time zero.

Nothing. Less than nothing, not even nothing. In a way, it was what I had come to see or hear. The absolute. On a Hollywood backlot, with all the scenery and the actors removed, between movies. As unimaginable as the universe itself. The opposite of imagination. The tabula rasa. Everything scrubbed clean. No thing, no body, no soul. No Being, no beings, no beetles, no Beatles, no bears, no bacteria. No God, no gods (where would you put them? There is no where, no when), no New York, no old York, no planets, no plans, no particles, no sun, no day, no night, no Milky Way, no life, no death. No ribs, no teeth, no skull. No jellyfish, no sea, no Earth, no heaven. No beginning and no end, much less middle. Where and for how long none of this is not happening, it is hard to say, for neither is there time nor space. No what, no why. No, nor language either. No yes, no no. Nought, *nada*, *niente*, negation. There was/is (is not?) nothing. But then – except there is no then, no there either – as Unc put it so memorably, there is nothing to stop there being something. We are in the realm of quantum uncertainty, when to be or not to be really is the question (and the answer is: both).

The thing that is no thing suddenly acquires pith and substance and the tide flows back outward, billowing out into space and time again. We are back before the beginning/after the beginning. Everything goes into reverse: we are watching the rise and the fall of the universe on rewind. Genesis is also Apocalypse. Every beginning is an end, birth seems to require death. Seen back to front, the extinction of everything in this alternative frame looks like the emergence of fresh life, an ending twisted around and inverted into a brand-new start. *Inflation* (in Alan Guth's theory) instantly takes hold, multiplying and being fruitful. Matter bulges outward, building, blossoming, ballooning, flaring up, igniting, exploding, spilling and spreading out, over and beyond, transcending itself again and again, unfolding, and putting out the deckchairs on the beach. Is it the same universe or another? Perhaps it is a mirror image, a mirror universe, reflecting all the elements of the first, but younger. Or, to think of it another way, every universe is a vulture feeding off the corpse of a dead one. Everything is recycled, nothing is wasted. Eternal recurrence. Another sun, another Earth, another Los Angeles. And so we return to where we started. You can turn it around, or inside out, rewind, fast forward, and it comes out the same, even though different.

A nine-year-old Canadian boy named David once sent me a short story about him and his dog. A true story. It ended like this:

THE AND

I think he had espoused the principles of the Big Bounce. No 'boundary conditions', no 'limit point'. The beginning and the end had all been swallowed up by an infinite middle.

And where is God in all this? the viewer of the DVD might ask, as Napoleon once asked of Laplace. Nowhere? Everywhere? 'A hypothesis I had no need of, sire.' Perhaps God is the viewer, since there is no other conceivable observer. Perhaps we are just looking at it on the wrong scale. All this tremendous compression and expansion – breathing in and breathing out – all this could be just the respiration of God, systole and diastole, a single heartbeat of one of the Slow Ones?

The Big Bounce had one major drawback, what the theorists at Penn State called 'cosmic forgetfulness'. It was like our previous lives, if we assumed reincarnation: it was hard – except in rare circumstances – to bring them back to mind. Same with previous universes. No footprints in the sand, no sand. No fossils, no evidence of any kind. On the other hand, the Big Bounce and the Quantum Loop definitely opened up the floodgates. There are 'free parameters'. Once you admitted more than one universe, there was really no stopping them from multiplying and proliferating. And each one of them was a little different to the others, none was a perfect replica of any other. 'The eternal recurrence of absolutely identical universes would seem to be prevented by the apparent existence of an intrinsic cosmic forgetfulness.'

It struck me that if you thought of these serial universes as simultaneous and parallel, then it was the 'many worlds' thesis, a natural development of quantum thinking. It was first formulated by Hugh Everett III in 1957 in *Reviews of Modern Physics*, although I think I first came across the idea in the pages of *Astounding Tales* or a novel by Philip K. Dick, in which the Nazis and the Japanese have occupied the United States. What if the wave function did not collapse (thus, so to speak, making its mind up, producing either/or)? What if Schrödinger's cat is alive in one version of the universe, but dead in another? To be *and* not to be. No more *or*. All alternatives are catered for. Everything bifurcated or 'decohered'. In this world I have white hair; in another world I would be the same, except with black hair (and my twin brother would have the white); in yet another I would be bald. In this world I wake up and get up and go to work and get hit by a bus; in another I am too hung-over or just plain lazy to be bothered to get out of bed. In this world I am married; in another I am Don Juan, or gay or a eunuch. In yet another I do not exist. In one world somewhere Adolf Hitler turns out to be a good guy who saves the planet from environmental pollution. In another the Atlantic slave trade is still alive and well. Napoleon won Waterloo and everybody speaks French and this book is called *Attention aux vaches invisibles*. We are all Neanderthals. Elvis lives (but can't sing). And so on. These 'relative states' (as Hugh Everett III – and, for all we know, I and II as well – called his alternate worlds) were like Einstein's 'inertial frames', but on a cosmic scale, and it was no longer possible for anyone to look over the shoulder of any of the others, any more than it was possible to peek into prior universes.

Driving back down Mount Wilson, I realized that if we assume infinite ramifying worlds in a schizoid 'multiverse', then fiction becomes impossible. Everything is true, somewhere, some time. Odysseus and the Cyclops, Don Quixote and the windmills, Sherlock Holmes and Philip Marlowe, Captain Nemo, the Hunchback of Notre Dame, and the Aleph, they are all out there. In one world, Emma Bovary is having unhappy affairs and committing suicide (or not). In another, a goofy-looking but friendly extraterrestrial is phoning home. That time I told my parents some kids had thrown a stone at me and broken the window – that would not be a lie. And then again, in some adjacent world, everything really is a lie. There is a universe with a benevolent, loving God; another ruled by Satan; and a third from which God and Satan alike are entirely absent, even though everyone believes in them anyway. All propositions and their exact opposites are true, there are no more falsehoods, no fantasies, no dreams, nothing that cannot be real. President Kennedy was not assassinated: he married Marilyn Monroe (with a best man by the name of Lee Harvey Oswald) and they lived happily ever after.

36

I was having lunch with Alan Weinstein. I had come back down to Earth (a very similar Earth if not completely identical) at Caltech. We were sitting outside the Chandler café, sunning ourselves and eating a big salad. And he was struggling to reconcile God and physics. He was also preoccupied with a bunch of hostile e-mails he had received that morning. Amid all the sun, the salads, and the leafiness, scientists at Caltech were living in a state of virtual war.

The e-mails were from people who didn't like what he was doing. There were a lot of people who fell into that category. It seemed strange to me, because I thought what Alan W. Einstein was doing was just about the most important thing in the world. But there were plenty of others out there who didn't feel the same way. Once it had been mainly Joe Weber, complaining that there was nothing wrong with resonant bars and interferometers were a fraud visited upon a gullible public, all LI and no GO. Now the antis were made up of two main constituencies: the fine upstanding tax-paying citizens; and the evangelicals. The citizens thought that their tax dollars could be better spent on just about anything else, if they really had to be. In fact, if you abolished interferometers and forgot all about gravitational waves, wouldn't it be possible to cut income tax? Really, who needed to know about the Big Bang and black holes and the like anyway? And we ought to come back down to Earth and concentrate on more immediate priorities.

They had a point. LIGO and all its relatives were not cheap. On the other hand, if you compared it with the large hadron collider in Geneva, it didn't look too expensive either: 'They have billions of dollars and thousands of physicists,' Alan said. 'And what they are looking for is much more exotic.' The funny thing is that as he was saying this, one of his graduate students passed by, a young woman, blonde and tanned, who had been working at CERN and was returning to base: 'D'you find the Higgs boson yet?' Alan enquired. The so-called 'God particle'. 'Nothing yet,' she replied amiably. 'We're hoping we'll come up with something nobody has thought of yet. We want to be *surprised*.'

The odd thing was that the other group of refuseniks made the exact opposite point. The fundamentalists thought that the origin of the universe was too important to be left in the uncertain hands of PhDs and postgrads. Creation was the proper reserve of God and God alone. Everything you could ever wish to know on the subject had already been spelt out by the book of Genesis. Only God could pretend to perfect omniscience and these physicists were arrogant enough to think they could know the mind of God. It was not far off what the native Polynesians thought about having the Keck sitting on top of Mauna Kea. Scientists were trespassing on sacred ground. The divine fiat lux was not to be queried by a bunch of guys in plastic booties and hairnets. There was a strange congruence between twenty-first-century fundamentalists and nineteenth-century positivists: they all thought that science had no right to stick its damn snout into the trough of either Genesis or Apocalypse. Science should concern itself with the middle ground, all the time and space in between. Wasn't that enough for all reasonable purposes? The trouble was, from Alan's point of view, that was just the part that was least appealing. 'There are only two interesting moments in history,' the French sociologist Jean Baudrillard once said, 'the Big Bang and the Apocalypse.' Everything else seemed relatively banal, derelict and tawdry by comparison. Alan Weinstein agreed with him. Physics couldn't hope to get much of a grip around the bulging waistline of the ouroboros, the natural habitat of human beings.

My respect for the Caltech crew increased in proportion to the amount of hostility they aroused. Nobody ever did anything original and groundbreaking without bumping up against the traditionalists. Kip Thorne – the Godfather – got most hate mail. He was like the icon and figurehead of the whole project, and therefore its number-one scapegoat. But Alan got his share, and so did everyone else. (I'm half expecting to get some myself.) They were all in the firing line. Their hands were dirty, either that or they were reaching straight into the taxpayer's back pocket and stealing their hard-earned cash. And then, wasn't their secret plan to prove that God is dead? Anywhere else they could be strung up or beheaded for blasphemy.

Again I felt that the hard-line fundamentalists (Christians mostly, with perhaps a sprinkling of Muslims, Jews, Shintoists, hippies and Neopagans) who were leading the fight had a legitimate point. Weren't all physicists godforsaken, god-hating atheists? Wasn't the whole gravitational relativistic exercise a little like an extension of Laplace: weren't they out to prove that God/Allah/Jehovah/Thor was a hypothesis that one could perfectly well do without? Superfluous, as Einstein had said of the concept of the luminiferous ether. If you could track the universe all the way back to the origin, then surely the interferometer would register the presence or the absence of any deity. The beginning, not the Flood, not the Exodus, not a burning bush or an empty tomb in Jerusalem: it was the most promising of all locations as regards pinpointing the Creator. 'Thou canst not see my face and live'. Thus spake Yahweh. But all those gravitational waves: wouldn't they finally reveal the true 'face of God' – or fingerprint, or footprint, or divine DNA or something? Or, on the contrary – and this was the fear – not? The point about the Genesis moment had to be that 'in the beginning', either the Earth was without form and void and the Lord moved upon the face of the deep and said Let there be light, or not.

And if not? Then didn't that just about wrap it up for theology? The great game would be over. All the priests and the mullahs and the Pope could all admit they got it wrong, apologize, and go home. It was – I had to admit – a tempting thought. It would be a radical simplification of the world, perhaps – who knows? – an opportunity for us to bury our differences once and for all. The light of pure knowledge would eclipse and obliterate mere belief. All those religious conflicts would be a thing of the past and could be viewed only on distant planets rather than, in extreme and painful close-up, on this one.

It was a beautiful theory. And some of the people sending e-mails to Alan definitely feared it could come true. It was like a war between science and religion and only one of them could win. Alan Weinstein was like the new Galileo, liable to be picked up and interrogated by the Inquisition – and worse. (Denying one faith could be enough to merit execution – but denying faith, all faiths? Mere execution would not be enough. How about sticking him in Schrödinger's box, along with the cat?)

Alan was a sceptic at heart. He had a few scathing things to say about the true believers who were writing to him so fervently and threatening him not only with hellfire and brimstone but also with making sure – by way of human intervention – he got his just desserts sooner rather than later. (The taxpayers only wanted him to be imprisoned for illegally flushing their money down the drain rather than actually tortured.)

And he was an optimist where the progress of knowledge was concerned. Even knowledge beyond physics. Of human behaviour, for example. 'In the twenty-first century complex systems will be better understood than they are now. Maybe in ninety-three years we'll be able to apply some of this stuff. We can bridge the gap.' He stopped to think about what he had said. He ran a hand through his hair. 'And then again, maybe not.' The more he thought about it, the more he thought that the gap was unbridgeable. 'Because it's a qualitative gap – and it's owned by the spiritual guys.' Alan wanted physics to succeed but he knew it was never going to explain everything. Unfortunately. 'If the laws of physics kept on changing we'd give up. But do the laws of human experience change? – yeah! It all depends on who you're married to.'

Then again, maybe there was a law in that too. Alan not only had a wife, he had a brother-in-law. And his brother-in-law was a Sufi master and deeply spiritual and spoke of the 'Six Subtleties' and 'states of the heart' and suchlike. The brother-in-law didn't send him e-mails, he didn't need to, he went ahead and told him face to face. '"Over here we have the godhead: this is huge, vast, incomparably important. And then there is everything else: it's all microscopic, only matter and energy. Applying the scientific method to the godhead just won't work. There is no fit."'

Alan believed in the Sherlock Holmes methodology of science: when you've taken away everything that is impossible, what remains must be the truth. Perhaps not the whole truth, but at least nothing but the truth. In a sense, he wanted to be like the TV physicist gumshoe, pursuing perpetrators and dark forces through space and time. 'Detectives are trying to discover the facts. And we're doing the same thing. You want nature to tell you the whole story honestly. You want it to be unbiased. And we don't want to bias ourselves either. We go into this eyes wide open. We would really like to discover something we didn't expect.' Despite all the angry e-mails from Outraged of Kansas City, Alan was really trying to be good, to do the right thing, and go through the narrow gate. But in the end, it was true, they were biased. 'We are so biased, we are willing to spend hundreds of millions of taxpayers' dollars, we are so certain these things exist.' They were biased in favour of truth, there was no way round that. 'We are biased: we like laws that are consistent, so we can infer from experiments. The godhead is not consistent. The godhead can do whatever it feels like doing. You only have to have a look at the Bible. Plagues, locusts, global annihilation, and then it's all sweetness and light – He's very moody.'

Alan believed in the Big Bang. It was a 'very big extrapolation'. But it was normal that they should extrapolate in the direction of the tail and the head of the ouroboros. 'Mathematics was invented to get us from what we know to what we don't know. Wigner [the physicist Eugene Wigner] used to speak of "the unreasonable effectiveness of mathematics" at describing the physical universe. So yes, it is speculative, but it isn't *idle* speculation. It's deeply informed speculation. You throw it out if it conflicts with what you know. The bedrock of science is testing. The point of LIGO is trying to put it to the test.'

But even when – *if*, for the benefit of sceptics – LIGO comes up with a wave-map of the Genesis moment, time zero (or 0 + Planck time, about 10^{-43} s, the physicists would say), it still isn't going to blow away all belief. The evangelical constituency didn't really have to worry, no mullah or archbishop was going to be put out of a job by LIGO. 'The godhead still wins,' Alan said, a bit peevishly. 'Anything we can come up with, my brother-in-law is going to say: that's still in the realm of maya. It isn't divine. The divine can always trump your hand.'

In the end, the strange thing is: Alan Weinstein didn't really have any objections to the godhead. He felt that the godhead and cosmology could coexist reasonably harmoniously. He was still a practising Jew. His enigmatic wife was Jewish. His son was due for his bar mitzvah soon. Alan thought that if he was a serious agnostic about the whole thing, he really ought to give up and wash his hands of religion once and for all. And it seemed as if he had been tempted a few times. But he didn't renounce. He still dropped by the synagogue from time to time. He still said Kaddish ('May his great name be blessed forever and to all eternity'). Maybe it was dyed into him from birth. But he found, when he came to consider the matter, that he couldn't readily dismiss the idea of a soul. He couldn't make any sense out of Yahweh, Yahweh was just too mysterious, but the soul: the mystery made human: that he could understand. Or rather not understand, that was the point of it. 'Don't you think you have a soul?' he said.

I'd never really thought about having a soul. 'If I have a soul,' I said, 'it must be the part of me that says, "I am *not* whatever it may be" – like surfer or scientist or accountant. I may be one or other of those things, but I also feel as if I could be anything or nothing. But isn't that just the same thing as consciousness?'

'Consciousness!' Alan exploded. 'But that is just *it*. And not so much my own, but other people's. My wife must have a soul, for example, which is what makes her so hard to predict.'

'Hell is other people,' I said.

'Jean-Paul Sartre?' he said. 'Yeah, I can identify with that.'

Sartre was a militant atheist. But I could see now, from Alan's Jewish existentialist point of view, that the same philosophy could turn around and point in almost the opposite direction. The same thing that made people like hell was also what made them, from time to time, like heaven, in some remote way bearing a resemblance to God. Likewise the fact that you never quite knew if they were going to be heaven or hell on any particular day. If nothing made any sense, then everything cancelled out and religion made as much sense as anything else in life.

Weinstein's thinking chimed in with the subversive alternative reading of the book of Genesis, the opening of the Torah in the Jewish conception. 'In the beginning . . .' says the very confident and assertive English translation (in the King James Bible), supporting the classic *creatio ex nihilo* interpretation of the moment of Creation. First nothing, then something, in that order. But that didn't square too well with the following sentence, the section to do with the *toho bohu*. 'The earth was void and without form and the spirit of God moved upon the face of the deep.' In the second sentence – the one after the beginning – it seems as if there is a lot of raw material already in place and Yahweh comes along and spins it up into something new and special and distinctive, shot through with goodness (and, therefore, potentially, evil). There are other translations of the opening phrase – *Bereshith* in the original – which make these dual perspectives more compatible. The first noun can be read as a genitive, giving something like 'In the beginning of creating . . .' In this version, the first main verb wouldn't even occur until 'God said . . .' Yet another alternative and perfectly valid reading would give: 'In *a* beginning . . .' There is no absolute beginning. Everything recurs over and over again, with variations. It was the Big Bounce rather than the Big Bang, eternal recurrence Nietzsche-style. By the same token, 'the' universe would turn out to be only 'a' universe, one among many, and not necessarily the best – or the worst. The raw matter and energy that the/a universe is made up of goes around and around, and the role of Yahweh is then limited to (a) observer or sentinel, (b) occasional interventionist. He is more of a gardener – perhaps a guardian – than someone who invents the soil in the first place. The most he can do is plant seeds and weed a little here and there.

And, for Alan Weinstein, perhaps the godhead was somehow bound up in the human soul. Unlike his brother-in-law, he didn't think that the godhead had much to do with the origin or the end, Genesis and Apocalypse, alpha and omega, the realm of the absolute. That was really the preserve of physics and mathematics. The divine really only has anything to do somewhere around the waist area of the ouroboros, at the same time as living beings arise, be they human or otherwise. Good and evil, God and Satan, the godhead and maya – they all belonged to the zone of indeterminacy, the same domain that his wife inhabited. It was the same world Alan Weinstein and I lived in. A world, as he might have put it, 'with hair'.

Arthur C. Clarke would have felt at home here. His great work 'The Sentinel', which would eventually turn into *2001: A Space Odyssey*, proposes alien intervention in the course of our evolution. Super-intelligent extraterrestrials plant a monolith on Earth which, responding to the presence of early hominids, artificially enhances their neural pathways to propel them on to the next stage of development. Isaac Asimov relied on various kinds of alien interaction somewhere along the line to steer and stimulate. In one story he asks the question: Where did jokes comes from? It is a version of the origin question. His three computer engineers realize that they are always saying 'I heard a good one the other day' but can never quite locate the absolute source of the joke. So they feed the question into a giant computer. The answer comes back: 'Aliens invented jokes and are passing them on to humans as part of a psychological experiment.' The three men have a good laugh about this absurd idea, but they decide nevertheless – humouring the computer – to ask a second question: What will happen if we find out what is going on? The answer comes back: 'The aliens will cease their experiment.' They roll along with the idea and now they have gone this far, they realize they have to ask a third and final question: What will the consequence be? The giant computer chews over this question for a while and finally spits out the answer: 'The aliens will show no further interest in

Earth, jokes will disappear, and humans will lose their sense of humour.' The last lines, as I remember them, are something like this: '"That has to be a joke," Bill said, "doesn't it?" But nobody laughed.'

The alien intervention device is always attractive, perhaps irresistible, to the science-fiction writer; it explains many things that otherwise we have difficulty in explaining – the origin of intelligence, of jokes, of anything and quite possibly everything. It also suggests why, in part, God is a hypothesis that will never entirely go away, why, as Nietzsche put it (when he was not saying 'God is dead'), slightly grudgingly, 'we shall never get rid of God'. Whatever the effects we are trying to analyse and understand, it is always possible to formulate another preceding cause that seems to wrap up and make sense of everything that comes after it, the First and Final Cause, in the shape of the extraterrestrial, the transcendent: stardust, the monolith, extremely intelligent aliens, or some force beyond mere aliens that puts us on the same branch as them and embraced by the same global power of causation.

I realized, somewhere in the middle of that lunch with Weinstein, that it was the very elusiveness of a foundation and origin, the fact that we arrived too late and missed the start of the show, that enabled or compelled us to dream of gods and aliens and primal atoms.

37

I had to go for a walk. I wandered away from campus, the cloisters and the cafés. It was the middle of the afternoon. The middle of Pasadena. I didn't have anywhere particular to go. I took a right and a left, a left or a right, walking at random. Everything was in bloom, bursting with colour and sap and nectar, cherry trees snowed under with blossom, jacaranda, mimosa, like billboards for spring in California. Skimming from flower to flower like a bee or butterfly, I ended up at the Huntington Museum. I'd heard about its wonderful collection of art. But when I got there the museum was on the verge of closing, so I didn't go in and strolled around the gardens instead. I had this feeling of being blown around on the breeze like a hayseed. I went through the rose garden and the Australian garden – full of eucalypts – and a bamboo forest that put me in mind of Tahiti, and then I came to the Zen garden.

The thing about the Zen garden is it didn't have any plants to speak of. In one way, it was the opposite of a garden. Maybe there were a couple of highly sculpted azaleas in it, I'm not sure. It was mostly rocks, beautiful, random rocks, just a few of them, artfully distributed around the space (perhaps 60 by 20 metres), and in the distance some kind of pagoda. But the base of it all was smaller stones, gravel, almost sand, the kind of tiny pebbles that have been washed clean by the tide advancing and reversing over them for a million years or so. It looked like a beach from which the ocean had long since receded. I was reminded of that key concept in Zen thought: *wu-hsin*, 'no-mind'. There was nothing to identify with in this garden, almost nothing to identify, no perfumes to permeate the head and the heart, no blooms, no stems or stamens, no host of golden daffodils. It was a garden of nothingness, a negative garden, an approximation to the tabula rasa. A state of rest.

And then I noticed the waves. Everything was flowing. The ocean had not really receded at all. Someone had raked waves into the stones, or rather raked the stones into waves, small but well-formed sine curves, rolling from one end of the garden to the other, cartoon caricatures of perfect waves, endless, like an image of eternity. I was put in mind of another poem of Wordsworth's – nothing to do with daffodils – 'Intimations of Immortality':

> Hence in a season of calm weather
> Though inland far we be,
> Our Souls have sight of that immortal sea
> Which brought us hither,
> Can in a moment travel thither,
> And see the Children sport upon the shore,
> And hear the mighty waters rolling evermore.

The Zen garden seemed like an overflow of that immortal sea. The poem and the garden exemplified something about physics and my own search for the soul of the universe. If everything is history, then all knowledge is recollection.

Now I was back on the West Coast, with Hawaii a few thousand miles further west, I couldn't help but be reminded that what got me started on this journey back to the beginning was the desire to escape from surfing. And yet surfing had been coming after me all the time. As Alan Weinstein said, you basically couldn't move without bumping into another wave. Thales (one of the early Greek philosophers who preceded Socrates) said, 'Everything is water.' Perhaps he was thinking that everything is made out of waves, not just the ocean but sound and light and gravity, that the whole universe tends towards the waveform, as if this were the basic template, the original archetype, the only way to travel across space.

I was standing staring at nothing when a venerable Zen gardener with a short white beard came up to me. 'Closing time,' he said. My first semi-insane thought was that he had found some way to close time, shut it right down, and thus achieve immortality. 'Soon,' he added.

38

'The first wave could be in there somewhere,' Professor Lazzarini said.

'And we would never know,' added Mr Anderson.

The problem of tracking down the beginning of things had just been doubled. I was inside a pocket universe which was a digital version of the macro-universe beyond. But the pocket version wasn't any easier to get to the bottom of than the big one it was supposed to be analysing.

This was one of two conversations I had had at Caltech that day that came floating back to me at the Zen garden.

Albert Lazzarini was deputy director of the operation. The director was off somewhere doing diplomatic things, trying to keep up political support for LIGO, maintaining the lifeline, until the truth finally manifested itself. Lazzarini was powerful enough to be relaxed and benevolent and put me in mind of an elegant old-school movie director like Frank Capra (who actually went to Caltech) with a tweed jacket and slacks, perhaps even a pipe (or a cigar). Except for one thing. I remember a scientist at Cambridge once defining a scientist as 'someone who wears socks with their sandals'. Albert Lazzarini was one of the few scientists I met at Caltech who fitted the bill. He was unlined and unstressed and genially confident about the imminent revelation. He introduced me to the man who would surely be the first to witness it in all its digital glory.

Stuart Anderson was tall, white, thin, and did an excellent impersonation of Agent Smith in the movie *The Matrix*. 'Tell me, Mr Anderson,' he would say, 'what good is a phone call when you cannot speak?' or 'Do you hear that, Mr Anderson? It is the sound of inevitability.' Which is why I can't help but call him *Mr* Anderson. Mr Anderson took me to 'the Matrix'. This is what he called it. That and the 'Truth Factory'. It was a large basement around the same size as baby LIGO or the burial chamber of a pyramid. It was solid silicon from floor to ceiling. I have never seen so much raw computing power in my life before.

A lot of the time I like to work at the computer lab in Cambridge. There's a room there with a bunch of Macs and not too many human beings that feels like the equivalent of a gym, but with computers instead of weights. A good place to work in. Every now and then I used to bump into the old IBM mainframe in the tower: it was like one of the classic machines imagined by Isaac Asimov, and would have been called 'Brainiac' or 'Multivac' or something like that, a close descendant of the first machines invented by Turing and others during the Second World War for decoding encrypted German messages. All these great old machines had one thing in common: they were huge. Asimov imagined them expanding to the size of a city, perhaps ultimately absorbing the entire planet. They were monsters of data magnitude. The bigger the brain the better. Then miniaturization came in and disaggregation and the vast machine was disassembled and broken up into fragments. The IBM (Brainiac) dissolved into PCs. And then, as time went on, the fragments were joined back up again into loosely assembled networks.

The Matrix was a reborn Brainiac with more finesse and speed. It took small but dense units (or 'nodes'), around the size of a DVD machine, and stacked them all together, like baguettes in a bakery or tiles in a mosaic or blocks in the Great Pyramid or cards in a deck, to produce an unimaginably powerful architecture, measured in petabytes (beyond gigabytes and terabytes). The Matrix was something like a million times faster and smarter, or maybe a billion, than the old Cambridge lab. It was the brain of the world.

'You know the library of Congress?' Mr Anderson said, with a note of pride. 'It's around 10 terabytes. Well, we've got a hundred of those in here.' Within the cold room there were great banks of sheer intelligence, stacks, and occasional independent towers or obelisks (I think Mr Anderson called them 'silos'), shaped like hexagons, the cells of a hive. There was a hushed, sussurrant atmosphere of quiet industry, like the wings of a thousand insects beating, as the hive went about its business of monumental number crunching.

I suddenly recalled why it was Unc gave up satellite astronomy. His doctoral thesis was dedicated to recording the discoveries of an X-ray satellite he had helped to put in orbit. 'The key thing,' Unc told me when he was writing it, 'is you must never say "I".' (I – sorry, Unc! – checked and he had even managed to leave it out of the preface: 'this thesis was written . . . this experiment was conducted . . .' It was like a journal that read, 'The bed was gotten out of, the bathroom was gone to, the stairs were gone down, and a cup of coffee was made.') His satellite was brilliant at discovering planets going around far-flung stars that were, hypothetically, capable of being inhabited (by, as they would say on *Star Trek*, 'life forms'). There were only two drawbacks: (1) Unc was expecting it to go into a spin and decay and crash and burn through the atmosphere some six months after launch, whereas in reality it kept on going relentlessly, and went on transmitting data all the way through Christmas and into the New Year. 'Isn't this bloody thing,' Unc said, 'ever going to fail?' (2) The amount of data it generated was huge. Unc (and his colleagues) were only able to sample a small amount of it. The paper the satellite generated filled a large room from floor to ceiling (literally). It was like something out of a Marx Brothers movie – they could hardly get into the room it was so full. Every now and then they would pull out a few pages and have a look. And the longer it stayed up in orbit the more data it spat out. The machine was actually too efficient, too fiendishly successful, for mere earthlings – the Slow Ones – to be able to process all the information.

Fortunately, in the era of LIGO and Mr Anderson's Matrix at Caltech, processing power had increased massively. The Caltech computer room sucked in vast amounts of data (hundreds of millions of pages, great pulped-up forests, if you were crazy enough to print it all out) from the interferometers at both the Hanford site and Louisiana. Some came straight down the line, untouched by human hand. Some of it was bundled up in discs and FedExed to California and plugged in by hand (two mailings a week). On top of that it was getting supplementary dispatches from associated institutions as disparate as Michigan and Glasgow and Cardiff, all on the lookout for wave bursts. And it was hooked into thousands of individuals PCs, which it would borrow for periods of the day, to enlarge its power, while they were on sleep mode (the so-called 'Einstein at home' system). Every day LIGO was observing – or tuning into – phenomena all across the sky and (potentially) all the way back to time zero.

According to Moore's Law, machine intelligence increases exponentially. Gordon E. Moore first noticed back in 1965 that the number of transistors on any integrated circuit for minimum component cost doubled every two years. So – fast forwarding to chip technology – the amount of data processing CPUs are capable of has similarly doubled every couple of years (some put this at 18 months rather than two years). Unc finished his PhD around the end of the seventies, call it 1980. Twenty-five years on, the capacity for data analysis had escalated (I calculated) by a factor of somewhere between 100,000 and a million. It was like inflation. Unc is frankly envious: he is like one of those footballers who used to be paid a pittance a quarter of a century ago, and if he were still playing today he would be earning a fortune. All the figures had gone up, as they say, astronomically. One physicist I had been in touch with, Frank Tipler at Tulane, reckoned that God was nothing other than one almighty computer that would attain omniscience and omnipotence (and omnipresence, by sending out intelligent spaceships) some time around the 'Omega Point' and would then (re-) create the universe backwards.

I wasn't quite bowing down, not yet anyway, but I was blown away by the power, literally. Huge fans were blowing out of every stack trying to cool the rear ends of all these microcomputers beavering away all day and all night. Separate free-standing fans were scattered about, all spinning away like windmills, puffing up piles of papers and perfectly capable of blowing up the skirts of any passing Marilyn Monroe (of whom there were none). The industrial fans were all well and good, but the computer system also depended on cooling by cold water being piped in and around the whole room. But what would happen if they ran out of cold water? Even though it was only April, Pasadena had already had three or four 100-degree (F) days in succession. Caltech had actually run out of cold water and the temperature in the truth factory shot up and the efficiency of the machines shot down. Truth was accessible, but only at the right temperatures. If you got too close to the sun, all the wax started melting and the whole thing came apart and you crashed. Hence the extra fans. In outer space, coolness was free; on Earth, in an era of global warming, coolness was getting more and more expensive, and everything was overheating proportionately, following some kind of insane counter-productive version of Moore's Law, more akin to Murphy's Law – given long enough, everything will eventually go wrong.

'How do you know if the computer has finally picked up a wave?' I said.

'What do you think we should have, Stuart?' said Lazzarini. 'An alarm of some sort? Balloons?'

Mr Anderson: 'How about some smoke coming out of the roof?'

'Way it's going,' Lazzarini said, 'looks like we're going to have smoke coming out of the roof anyway.'

39

I came to the end of the Zen garden (not that it had an end). There was a miniature forest of bonsai trees. Each tree was a shrunken knot of gnarled and twisted trunk, and clusters of leaves, fluttering and catching the light. I was looking at the trees but I was thinking of what Sterl Phinney had said about what the Big Bang sounded like.

Sterl Phinney had been at Cambridge as an astronomy PhD student and we knew a lot of the same places and a few of the same people. He even used to give the kind of guided tour around the telescopes for interested amateurs that I had been on in the past, so – somewhere along the line – he had probably taught me a few things already. When he went to Caltech he had been given the job of floating LISA. He had spent years writing proposals and selling it to senators and generally putting bows around the idea of an interferometer in space. As far as he was concerned the project spoke for itself and, several years down the line, he had begun to feel that shuttling to and from Washington was not his idea of fun and he was scaling down all the propaganda and diplomacy. A pilot LISA craft was due to be taking off in the coming months, and the engineering was going ahead, but it was still likely to be years before the finished product went into orbit around the sun. And it was running into competition from NASA's upcoming Mars mission.

The point about the Matrix, Sterl said, was that it had to know what it was looking for in order to find it. All that data was useless unless it had some kind of shape to it. In the end, it was just a bunch of numbers. Numbers needed interpreting. Therefore the machine had to have a template, a grid, an identikit picture of the perpetrators it was pursuing through time and space. The simple but gruelling task all that massive computing power was carrying out was comparing one column of numbers ('gravitational wave strain data') with another column ('template signals'), like a pair of stereo speakers, and waiting for a match-up. It was one of the classic tensions in all science: you wanted answers but you had to be able to ask the right questions, so you came up with sketches, conjectures, hypotheses about the way things are.

Sterl, for example, had already worked out what an inspiral (like the Hulse–Taylor, with a neutron star swinging towards a pulsar) would be like. There were visual forms of an inspiral. You could probably do a video with decent special effects. But Sterl didn't draw me any pictures and he didn't show me any movies. Instead he sang to me like a troubadour of black holes. He could easily whistle up a supernova. An inspiral was more like an extended chirp, like the song of an uninspired nightingale. He apologized for the inadequacy of his rendering. 'This is a very speeded-up version,' he explained. 'In reality, you'd only get this kind of signal over a period, a year or two. I'm giving you the edited highlights, the grand outlines.'

Everything that happened in the universe could be rendered in the form of sound. It was the scientifically updated version of the music of the spheres or *musica universalis*, the classical concept of the harmony with which the celestial bodies wheeled across the heavens. Pythagoras didn't literally think you could hear the motions of the sun and the moon; but Sterl saw no reason why not. Everything you saw or heard was just a set of signals, correctly interpreted. I knew about white noise, the static that is also the sound of the remnants of the Big Bang several hundred thousand years after the Genesis moment, the first emergence of photons. I could hear it on my radio between stations. But I knew nothing of 'brown noise'. Sterl explained that it was any series of random notes, without any key, something like a composition by Schoenberg or the sound of waves breaking on the shore or a particle – if you could hear it – undergoing Brownian motion. Mozart, stars, and heartbeat rhythms were more in the realm of 'pink noise': notes strung together into a series but nevertheless governed by certain clear relationships. Sterl had even written – or rather got his computer to write – a piece of music based on the concept of pink noise (he didn't think Mozart would feel too threatened by it though).

The kind of template signals Sterl mentioned specifically were 'Harpsichord playing chromatic scale', 'Person saying "Like a Virgin"', and 'Prediction of general relativity for merger of two 1.4 solar mass neutron stars'.

'What do you reckon the Big Bang sounds like then?' I said.

I knew it was a stupid question. It was a bit like the question of whether or not, if a tree falls down in a forest and there is no one around to hear, it can be said to make a sound at all. But more so. The Big Bang, of course, didn't make a bang, not even a small one. It couldn't: you can't hear anything in a vacuum, sound needs a medium to propagate. And in any case, as in that mythic forest, there was no one around to hear it anyway. As the tag line for *Alien* put it: 'In space, nobody can hear you scream.' Even less so when there is no space, and no time either.

But Sterl didn't think it was stupid. He took it entirely seriously.

The point about the beginning of the universe is that, needless to say, humans did not exist at the time to perceive it. There were no official records of it taking place, no eyewitness reports, no cellphone photos, no YouTube videos. And yet what is also true is this: we are contemporary with the beginning of the universe. We really can hear the mighty waters rolling evermore. The Big Bang is still happening, it never finished. Nothing ever really finishes, everything still exists in the form of a signal, therefore it exists. In some frame of reference (just not one that I personally can access) my father is still alive, and my grandfathers and my grandmothers, and their grandparents, and theirs, and every other soul all the way back to the point where the meaning of the word 'begat' is mysterious and obscure.

'The Big Bang is red noise,' he said.

'Could you hum it for me?'

'We haven't got it up on the website.'

'What's it sound like though?'

'It's all the notes – being played simultaneously. The whole keyboard. Every single last note, they're all there, they have to be, going off at the same time.'

'And it's still playing?'

'Never stops. It's there all right. We just haven't managed to hear it yet is all.'

Weber had imagined he could hear it. As I walked about the Zen garden, admiring the configuration of stones, and gazing at the wave pattern flowing through it, timelessly, like a hieroglyphic inscription on a pyramid, I had been trying to work out what red noise would sound like, I tried to hear it in my head, as if that would bring me closer to the absolute, and give me a fix on the origin. But all I could hear was the ghostly tinkle of wind chimes hanging in the breeze.

40

It was a dismal summer in England. We had become used to long hot dry summers, and drought conditions with hosepipe bans. This time around it was the exact opposite and half the country had been flooded. First it was Yorkshire, in the north, then it was the West Country and the middle of the country. Oxford was one of the places worst affected. I went to see Unc, who lived on – now striking a rather sinister note – Riverside Road. Oxford looked more like Venice than Oxford, a place of canals and mad, swirling torrents. At one end of Unc's street was a playing field that had turned into a lake and instead of kicking a ball about people were practising their kayak manoeuvres on it; at the other end, the main road to Botley was under two or three feet of water. Almost miraculously, Unc's house and a few others on slightly higher land were preserved, like an atoll, or an ark, amid the waters. He had spent a few sleepless nights though, listening out for the sound of water slipping in through the back door and creeping up the stairs like a burglar.

But whereas everyone else in Oxford spoke of sandbags and the inevitable watery doom that awaited them, as if they were passengers on the *Titanic*, Unc was excited by a brand new synchrotron that had just been built and activated at the Oxford science park. It looked like a baby hadron collider. There was a massive track, with huge 'wiggler' magnets wrapped around it at regular intervals, with indecisive quantum particles racing along it at speeds approaching the speed of light. The particles wanted to radiate out and the magnets kept channelling them back in again, so they wouldn't scatter. At one end of it there was a slip road, where X-ray particles were siphoned off and slammed through a microscope the size of a car. If you put your sample under the beam and looked through the lens (so to speak – the image is conveniently displayed on a screen in another room) you could see the fundamental molecular structure of any material. It was on the order of a thousand times more powerful than the typical electron microscope. Chemists could go about inventing new polymers and crystals and membranes, sticking molecules together in hopeful combinations, and then poke them under the synchrotron lens and see what new symmetry they had come up with.

'You've never been able to see this stuff before,' Unc said. 'You can actually see the way the atoms slot together.'

I felt a kind of pang. There were no more excuses. You could finally see the moonbeams, nothing was too small any more. At the same time, the synchrotron detector reminded me of everything that continued to elude me. I wanted to see the underlying structure too, the architecture, but not so much of brand new things that somebody had just invented as of the very old things that just came to be, the stuff out of which all other new things were ultimately made. There is nothing new under the sun, as the Ecclesiast said. Everything could be traced backwards towards the point of origin.

Unc was also speaking of the 'transactional interpretation' (of quantum mechanics, derived from the work of John Cramer) over breakfast. It explained how entangled particles got to be entangled in the first place. According to Unc's version of transactionalism, the two particles really were communicating with one another, but they weren't mystically linked in space, it was just that they sent out waves back through time (Unc called them 'Wheeler–Feynman waves') which ultimately caught up with one another, shook hands, and then flowed forwards in time again, enabling them to synchronize with apparent simultaneity. If you could make a movie of photons interacting, it would look the same on fast forward or rewind. These disparate particles were constantly in touch with one another by virtue of feeling their way back into the past. 'You can't prove it one way or another,' Unc said, buttering his toast. 'That's what "interpretation" means. But it makes sense.'

'Does it?' I said.

He tut-tutted at that and went and got a sheet of paper and drew a sketch:

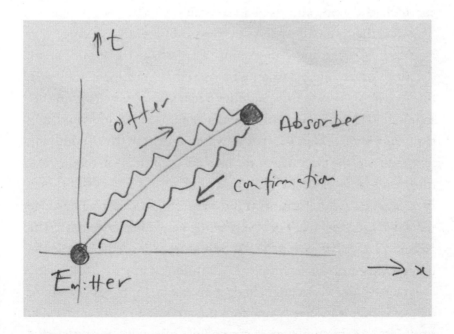

But it wasn't just particles, it was us too. When we stand in the darkness gazing at a distant star, not only are 'retarded' light waves ('offer waves') reaching our eyes, but we (which is to say, the electrons in our eyes) have been sending out 'advanced' (or 'confirmation') waves to meet them halfway and thus produce the 'transaction' that is our image of the star. A two-way exchange between emitter and receiver (or absorber) is taking place. A standing wave in spacetime connects us up to the stars. It was a strange kind of conversation to be having over toast and a bowl of cornflakes. Another two-way exchange between emitter and receiver.

By virtue of the same style of thinking, Unc felt completely relaxed about the idea of there being something and nothing at the same time. 'There's no big difference,' he said. The idea was built into quantum mechanics, the principle of and/and as opposed to either/or. It was a fuzzy logic, perfectly capable of embracing all paradoxes. Particles were waves and waves were particles. As soon as you had nothing you also had something. It was like the kind of particles that popped up in super-fast colliders all the time. That was one of the things that particle-watchers were particularly looking out for. They flashed into existence, glittered tantalizingly for an instant or two, bumped into a few other particles, and then blinked out again, as evanescent as butterflies, too fragile to survive for very long, and yet they had the strength to come into existence in the first place, for no very good reason, except that they could.

It was Saturday evening and we were on our way to an organ recital at Queens' College. 'Maybe the universe is like that,' Unc speculated as we crossed over another bridge into the aquatic city. On another scale, in a completely different frame of reference, in a super-collider viewed by aliens who would appear as gods to us, our universe was simply another stray, random particle that had spun up into existence and – from this infinitely more extended point of view – would simply burn out and die almost before its existence could be registered and properly studied. It was a mystery that could not be fully analysed and understood. An enigma. We were so small and so lightning-fast that the extremely large, long-lived and slow-moving alien life forms, who hardly had time to catch their breath while our entire universe was born, lived and died, would be incapable of working us out or even detecting our pointless, quirky, ephemeral lives. 'A new fundamental particle?' some alternative-universe postgrad spectator might be saying, before the lights went out again. And the spectator might wonder, too, if this microscopic speck of nothing could be entangled with some other pint-sized universe somewhere else, and they were sending out waves to one another, waves washing backwards and forwards through time.

Or perhaps no one would even notice as we coughed, spluttered, and faded into black.

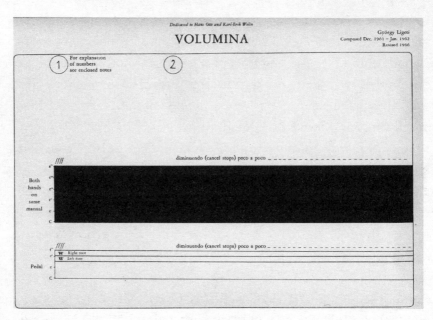

The Big Bang?

41

I was in Oxford mainly to pick up my older son, Spencer. If Unc was right, we were in constant touch via some kind of transactional wave arrangement, but if I wanted him to come home I still had to go over there and get him. He was now sixteen and a music scholar and he had been attending a gruelling week-long organ course at one of the colleges. We drove back to Cambridge, stopping off in Coventry to pick up a new puppy, only eight weeks old, with a lot of black curly hair, called Waffle. Spud spent the next month or so playing with the puppy, teaching him commands, and playing the organ (and other assorted instruments) and going along to recitals his friends were giving.

When I was in Cambridge, I loved to cycle off to his school early in the morning, before many other people were around, to hear him practise some piece of Bach or Brahms or Rutter. But, following my obsessional pursuit of the origin, and doing some teaching in New York, I hadn't heard him in a while. He told me he had been working on a new piece, which he described as 'very interpretive'. It was still 'work in progress', he emphasized, it was still probably 'too difficult' for him, but he had heard about it in Oxford and somebody had given him the music and he wanted to try it out.

I parked my bike next to the chapel. The school was a Methodist foundation created in the nineteenth century by evangelical Victorians, convinced of the virtue of good works. James Hilton drew on his experiences here for his novel about a dedicated, almost heroic schoolteacher, *Goodbye, Mr. Chips*. The chapel was simple and beautiful, a place of stone, pillars and stained-glass depictions of saints. Inscriptions paid tribute to boys and masters who had gone off to war and not returned. The organ was positioned next to the altar, opposite the choir stalls. It filled the whole chapel with good vibrations. We used to joke that this was why Spud had originally taken up the organ – at the age of around six or seven – because it was so powerful it was like the voice of God. And it had lots of knobs to play with, buttons and stops, not to mention three manuals, and a dozen pedals. There were so many controls it was almost like sitting in the cockpit of a plane or a space rocket.

He was in the middle of the piece when I sat down and listened. Unusually, he had his younger brother Jack sitting next to him and joining in. It was nothing like Bach or Brahms. The music sounded like some kind of riddle with no possible solution. There was no clear line that I could distinguish, no melody, and no harmony either, but a blizzard of notes, flying at me, as if I was driving through a storm. I got up and looked over their shoulders at the music.

'Look at the last page,' Spud said, turning the page over, still playing and stabbing at buttons. 'Unbelievable.'

It was a single solid black line, filling the stave, running across the page horizontally, as if painted on with a thick brush. A great wall of sound was coming from the pipes, sustained and unvarying and incomprehensible. A trumpet blast but as if an entire orchestra were belting it out. Spud was not playing with his hands any more, he was playing with his arms. He had them splayed out sideways so that his fingers were touching and he was hitting as many keys as possible. Jack was working all the remaining keys. Between them, they were depressing all the keys simultaneously.

I realized I was hearing red noise, great waves of it washing over me, the music of the beginning, the organ of the origin.

I read the directions on the score: 'Continue to sustain keys until the last pipe has ceased speaking and the last whiff of wind has disappeared.' It was 'Volumina' by György Ligeti (the name having an uncanny affinity with LIGO, as if he would have to be LIGO's court composer, and translating as something like 'little LIGOs', a diminutive on the model of *libretti* or *spaghetti*). At the bottom of the page, underneath the thick black line, was an empty stave, a rectangular box with nothing in it, a blank. The keyboard was doing everything, the pedals nothing. The final direction was: 'Observe period of silence lasting appr. 30 seconds, still continuing to depress keys.' It sounded more like the end than the beginning (although the same thick black line recurs, I later discovered, on the first page, so there was a continuum, a Big Bounce effect).

It was the kind of music where I was tempted to say (although I didn't say it), 'Hey, I could do that!' And, other things being equal, not only I but anyone else could too. Perhaps, it occurred to me, this was the music of the spheres, the *musica universalis* that was running through everyone's head like a jingle you can't get rid of, cosmic background noise, the ultimate wind chime, it was the sound of everything and nothing. The soul of the universe. It was like a small earthquake going off. Or mighty waters rumbling.

I hadn't seen the beginning, but I had heard it.

I wondered if those delicately balanced beams of light back at LIGO could be picking this up and Mike Landry was puzzling over how to eradicate yet another rogue signal.

42

Summer, the real summer, had finally kicked in. It was one of those long warm evenings in August and I had just finished a game of tennis with Jack on the grass courts at school. He was fourteen. He was tall with fair hair and blue eyes and looked completely unrelated to the Shah's widow. It was a little like looking into my reflection in a mirror, but with a time delay of decades rather than nano-seconds. When he was three, he said: 'I can remember being in heaven before I was born.' Now his favourite phrase of the moment was 'I beg to differ.' We were lying on our backs on the grass staring up into the twilight sky, getting our breath back, just lying there, lazing, emitting and absorbing.

The stars were starting to pop out. We got to talking about the Arthur C. Clarke story 'The Nine Billion Names of God'. Clarke imagines a computer engineer being called off to Tibet to fix a Mark V computer belonging to some Buddhist monks in a lamasery in a Himalayan valley something like Shangri-La (dreamed up by James Hilton, again, in *Lost Horizon*). The purpose of the automatic sequence computer is to keep generating nine-letter permutations of the Sanskrit alphabet (there are certain constraints: no letter can be repeated consecutively more than three times, for example), of which there are supposed to be some nine billion possibilities. The monks believe that each of these words is one of the names of God and that when all the names have been uttered (even if only by a computer) then the purpose of humanity and indeed the universe at large will have been fulfilled and it will all quietly fold and come to an end. The engineer thinks it's all gibberish but humours them – they are paying customers after all – and having fixed the computer finally goes on his way, being carried by yak over the pass to Kathmandu. Night falls and he checks his watch and realizes that it should be around now that the computer will have finished working out the nine billionth permutation of all the letters. He chuckles to himself at the absurdity of the whole venture. Still, he reasons, if that's what they want to do with their time and money, so be it. As he looks up into the night sky, he sees that, quietly, without any great fanfare, without a bang or a whimper, all the stars are starting to flicker and go out. The end.

'This is the exact opposite,' said Jack. 'The stars are just lighting up.'

They were flicking on all over the sky, as if somebody were throwing the switch on the Eiffel Tower. It wouldn't make a good end to a story, but in its own rather familiar way it was just as strange as the end of the world, possibly even stranger. Something rather than nothing. They were sending out their waves and we were sending out ours. The And.

I was staring in the direction of infinity, not looking at anything in particular. 'You know if you could see far enough,' I said, following my usual train of thought, 'you could see the beginning of the world.'

'Why can't you then?'

'Our eyes just aren't good enough. Or our ears. Or something. Unless yours are any better.'

Jack focused on infinity too.

'Can you see it?' I said.

He scanned the sky carefully. I wondered if maybe the next generation could be better at this kind of thing. There were still a few dark clouds here and there. 'Maybe there never was a beginning,' he said.

'How do you work that out?'

'It's like all those old fairy tales. "Once upon a time . . ." You have to start somewhere.'

'So we're back to the beginning then.'

'I beg to differ.' He was a bit of a young Jean-Paul Sartre.

'What about the end then? You've got to have an end.' I think I was starting to sound a bit desperate.

'No,' he said ruthlessly. 'No end.'

'No Hollywood endings?'

'I'm just wondering: do you think there's a way of seeing into the future – you know, if you had a good enough telescope and all that? Couldn't you win a fortune betting on whatever's happening tomorrow?'

I had a smart reply already forming. It was going to be something like, 'I already know the future. Everybody dies.' But it curled up and died somewhere in my throat and never quite made it to my lips.

No beginnings, no endings. No middle either, therefore. But what if – and here I was following Jack's thinking – it was the exact opposite? What if it was all beginnings and endings? I thought of all the things I could say to Jack and I choked up with the possibilities like someone in a supermarket haunted and stymied by all the different flavours of toothpaste.

I twisted my head around and looked at Jack. He was still staring at the sky. 'Maybe I can invent a time machine,' he was saying.

For no good reason, I thought of Alan Weinstein. Sitting in a bath. He had gone to a cosmology conference, he was giving a paper on gravitational waves and the shape of the universe. The conference was being held in a grand hotel in New York. It turned out that at the exact same time, the Society of American Cosmetologists were holding their conference in the very same hotel. Cosmologist and cosmetologist met and mingled like stray particles colliding in a heightened state. Waves interacting. In the evening Alan hopped into the hotel jacuzzi to unwind. It was a big hot tub. Soon he was joined by half a dozen extremely beautiful women with perfect complexions. He had an idea they couldn't be cosmologists. Now their waves really were interacting, right there in the tub. Alan thought of trying to talk to one or more of them. But he guessed that he might struggle to 'make time' (in his words) with a cosmetologist. So he lay back and wallowed in the bubbles and tuned in to the conversation.

He was amazed. They were talking like scientists. There was a whole science of cosmetology previously unknown to him. Stuff about chemical reactions and subcutaneous zones, electricity and light therapy. Theories about ways of stopping time. They had creams and potions called 'Immortel'. He had a feeling that they had stolen his clothes and there was some sort of minor intellectual crime or misdemeanour being committed. But since he wasn't wearing too much in the way of clothes at the time, he softened, and thought what the hell. They can be scientists too. Maybe everyone was a scientist one way or another. It was just that some of them were more attractive than others.

Cosmologist and cosmetologist – what was the difference? The letters ET, it occurred to me, were the answer, oddly enough. Alien and yet bathing, swirling around, in the same hot tub.

Scraps of conversation were swirling around in the bubble bath of my brain. Something Fred Raab had said to me about emitters and receivers. Something Unc had said the last time I saw him. 'It's surprising there aren't more universes being formed all the time. We must have hit an equilibrium.'

Jack was right. There couldn't be an origin, not in the sense of a beautiful, perfectly formed singularity, the primeval atom, identical to itself, a state of absolute oneness. The minimum had to be a duality. An entanglement, like a conversation, between two states, that were the same and yet different. Something and nothing. Heaven and hell. Particle and antiparticle. Information and entropy. It was the quantum logic at work, splitting and bifurcating. Out of those two things, like antithetical twins, came a third, something else, almost nothing, which was the mismatch between them. And the whole world sprang up out of that disequilibrium, this world and others like and unlike it.

No beginning of the world, no end, no burning bush, no *Primum Mobile* nor Elysium particle. And yet it was the closest I had ever come to being in the presence of God. All along I had been looking right at the Aleph, the soul of the universe, looking and not seeing. I saw, lying there on the grass, that God resided in the interstices, the intersections, the transactions, the twilight zone of crossed lines and coincidences. God is just the name we give to a lack of information. God is an infinite number of possible beginnings and endings. I had a feeling then that there is nothing that is not sacred. Everything was interesting, and not just the great perfections of the Big Bang and the Apocalypse. Creation was taking place at every instant, with whole universes swerving off through different gates down separate pathways, depending on a touch here, a word there. Everyone, everything, was a mini-interferometer, picking up rippling signals about the evolution of time and space, and transmitting them right back out. Every single sentence cocooned an end and a beginning. Genesis now.

It would make a big difference what I said back to Jack. Something like the fate of the universe hung on it, or at least *a* universe.

'I don't know,' I said. I had forgotten what the question was but it seemed like the right answer anyway.

Jack went off and wrote a song, 'What Will Become of Me?', that ends with the words 'The future's just begun. You cannot underestimate the future. (rpt)'

I remember seeing a couple of digital clocks in Sydney, Australia, that recorded not time but deaths and births. The number of people dying was ticking around at the rate of nearly one a second, but the number of people being born managed to stay just ahead of it, by some significant fraction. I had a feeling universes were a little like that too, with vast multitudes fading out and going up in smoke all the time, and all the stars popping out once and for all, like bubbles, while at the very same time, or in an alternate time, new ones were igniting and flaring up, rising up out of the confusion and chaos, some better than our own, others infinitely worse, but collectively managing to keep slightly ahead of the great nothingness that threatens to annihilate them.

43

At least I knew now how to resolve one abiding mystery. All I had to do was get myself to a planet 54 light years away, borrow an extremely powerful telescope, and train it on Earth. Then at last I would see the moment of my own birth. And whether or not the widow of the Shah of Iran was there at the time.

The 'twin paradox' says that if one twin takes off in a close-to-light-speed flying saucer and finally returns to Earth he will find that, since his clock has been running slower, he is considerably younger than the twin who remained behind. I never quite made it into outer space and I think Unc looks older than me anyhow (and, in a relativistic way, he probably thinks the same about me). But I asked him to calculate the net effect of all the terrestrial journeys I had undertaken looking for answers. I was born second and came along ten minutes behind Unc, but the gap just got bigger. Assuming some rounded-up figures, I am now roughly another ten nanoseconds younger than Unc. His personal clock is ten nanoseconds ahead of mine. That is about as long as it takes for light to leave his face, bounce off the mirror in the hallway, and make it back into his eye. I have become Unc's honorary reflection. With everything reversed. Perhaps I always was.

The last time I saw Unc he was working on an anti-gravity device. He thought he might be able to get it to work on particles, but he had a vision of people ultimately flying weightless through the air, like photons, like moonbeams.

Through a glass, darkly, I can still see the first scientific experiment we ever carried out together. We were around six years old at the time. Somehow we had been impressed by the idea of sound waves. So we cleaned out a couple of cans (almost certainly Heinz baked beans), dug holes in the base, and knotted a long piece of string between them. We went into separate rooms in our house and stuck the cans to our mouths. We didn't know then that you had to keep the string in a straight line for it to work. And we never quite worked out who was going to speak first either, can to mouth or can to ear. The conversation, if you can call it that, went something like this:

A: 'Hello, are you receiving me? Over.'

U: 'Hello, hello, hello?'

A: 'Can you hear me?'

U: 'Is there anyone out there?'

We were both sending out waves but there was not too much in the way of transaction.

Which is how we came up with the first sketchy version of our string theory of the universe. Time and space are an infinite piece of string. We are all holding up empty tin cans to our ears.

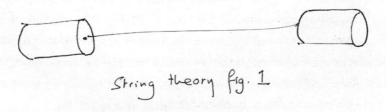

String theory fig. 1

There is one other involuntary experiment that took place when we were about three. It is one of my first memories, but perfectly clear. We were playing outside in the street. It was in Forest Gate, East London, across the road from Upton Park. I was racing along the pavement on a tricycle chasing my twin brother when I lost control and swerved off the kerb. The trike dived down into the abyss and so did I. I landed with my teeth wrapped around the handlebars. An ambulance – I can see the red cross – turned up and I was rushed off to the children's hospital nearby where they patched me up and stuck me back together again.

Everybody tells me that, in reality, it was Unc who was riding the trike, it was Unc who had the accident, it was Unc who went to hospital, and that I have got the whole episode back to front. But if that is true, then how did I get the chip in my front tooth? I suspect that alternate universes collided around us and there was some kind of interference effect. Perhaps we really were switched. One for the other.

Obviously, in some odd way, we must be entangled, Unc and I. Perhaps, in ways we cannot yet understand, we all are.

Every now and then, when I am in Oxford, I see Unc's old farcom, the flying-saucer phone. Bits of it lying about the place, a laser, the back wheel of a motor scooter, a few bricks. Mirrors. Remnants scattered around his house, on the wall, in cupboards, on shelves, even in the back garden. Just sitting there dumb and inert. But somehow they are still connected. They don't need wires, only light. Wherever I see light reflected, bouncing off the leaves of a tree, the feathers of a bird flashing by, the silver surface of a river, the white crest of a breaking wave, stones on a beach, fish, clouds, the moon, mirrors, Mars, Io, the mysterious face of a stranger passing in the street, it is hard not to see them as part of some immense, insane experiment, all connected up to an infinite farcom, beaming out obscure messages to one another.

I am still waiting for that phone to ring.

SOURCES

It would be like trying to name all the stars in the firmament to come up with a proper bibliography for this book. So I am only going to mention a few of the celestial objects that shine particularly bright in my mind.

Marcia Bartusiak, *Einstein's Unfinished Symphony* (Washington, 2000)

Martin Bojowald, 'What Happened Before the Big Bang', *Nature Physics* (August 2007)

Jorge Luis Borges, *The Aleph and Other Stories 1933–1969* (London, 1971)

Max Born, *The Born–Einstein Letters* (London, 1971)

Albert Camus, *Notebooks* (Paris, 1962–89)

Arthur C. Clarke, *The Nine Billion Names of God* (New York, 1967)

Harry Collins, *Gravity's Shadow* (Chicago, 2004)

John G. Cramer, 'The Transactional Interpretation of Quantum Mechanics', Reviews of Modern Physics, 58, 647-688, July 1986, also at www.npl.washington.edu/ti/

Albert Einstein, *The Collected Papers of Albert Einstein* (Princeton, 1987)

Hugh Everett, '"Relative State" Formulation of Quantum Mechanics', *Reviews of Modern Physics*, vol. 29, No. 3, July, 1957, pp. 454-462, and at: www.chat.ru/~everettian/english/paper1957.html

Paul Feyerabend, *Against Method* (London, 1975)

Robert Forward, *Dragon's Egg* (New York, 1980)

——, 'Wideband Laser-Interferometer Gravitational-Radiation Experiment' (*Physical Review*, D17, pp. 379–90, 1978)

Brian Greene, *The Elegant Universe* (New York, 1999)

Alan Guth, *The Inflationary Universe: the quest for a new theory of cosmic origins* (London, 1997)

Stephen Hawking, *A Brief History of Time* (London, 1988)

Walter Isaacson, *Einstein: His Life and Universe* (London, 2007)

Michio Kaku, *Parallel Worlds* (New York, 2005)

Martin Rees, *Before the Beginning* (London, 1997)

Richard Rorty, *Philosophy and the Mirror of Nature* (Oxford, 1980)

Edward Said, *Beginnings: Intention and Method* (New York, 1975)

Jean-Paul Sartre, *Being and Nothingness* (Paris, 1943)

——, *Nausea* (Paris, 1938)

Bernard Schutz, *Gravity from the ground up* (Cambridge, 2003)

Simon Singh, *Big Bang* (London, 2004)

Lee Smolin, *The Trouble with Physics* (London, 2007)

Dava Sobel, *Longitude* (London, 1996)

Kip Thorne, *Black Holes and Time Warps* (New York, 1994)

Joseph Weber, *General Relativity and Gravitational Waves* (New York, 1961)

——, 'Evidence for Discovery of Gravitational Radiation' (*Physical Review Letters*, Volume 22, Number 24, 16 June 1969)

Andy Martin was born in London and studied at Sussex, Cambridge, and the Ecole Normale Supérieure in Paris. His first book was *The Knowledge of Ignorance*. He teaches at Cambridge University and writes about waves, big and small. He is married to an Australian and they have two sons. Find out more at www.andymartinthewriter.com